# The Michigan Historical Reprint Series

*Reprints from the collection of the University of Michigan University Library*

This volume is produced from digital images created through the University of Michigan University Library's preservation reformatting program. The Library seeks to preserve the intellectual content of items in a manner that facilitates and promotes a variety of uses. The digital reformatting process results in an electronic version of the text that can both be accessed online and used to create new print copies. This book and thousands of others can be found in the digital collections of the University of Michigan Library. The University Library also understands and values the utility of print, and makes reprints available through its Scholarly Publishing Office.

For access to the University of Michigan Library's digital collections, please see http://www.lib.umich.edu.

The Scholarly Publishing Office seeks to disseminate high-quality, cost-effective scholarly content through both print and electronic publishing. Information about the Scholarly Publishing Office can be found at http://spo.umdl.umich.edu.

The Scholarly Publishing Office
the University of Michigan
University Library

AMERICAN HIGH-PRESSURE ENGINE

*Eight horse power.* *8 inch Cylinder, 2½ feet Stroke.*

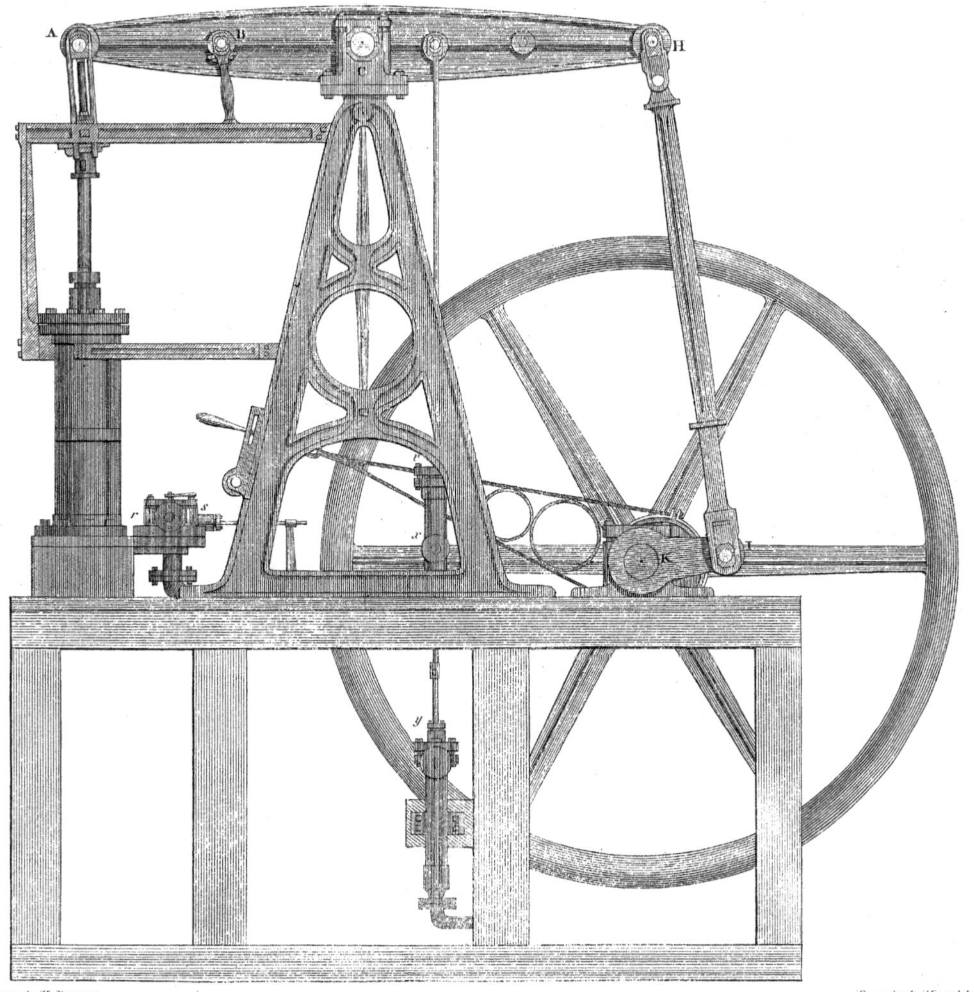


*Drawn by H. Brevoort.* *Engr. by P. Maverick.*

# THE
# STEAM ENGINE

## FAMILIARLY EXPLAINED AND ILLUSTRATED:

WITH AN

## HISTORICAL SKETCH

OF ITS INVENTION AND PROGRESSIVE IMPROVEMENT;

ITS APPLICATIONS TO

## NAVIGATION AND RAILWAYS;

WITH

## PLAIN MAXIMS FOR RAILWAY SPECULATORS.

BY THE

REV. DIONYSIUS LARDNER, LL.D. F.R.S.

FELLOW OF THE ROYAL SOCIETY OF EDINBURGH; OF THE ROYAL IRISH ACADEMY; OF THE ROYAL ASTRONOMICAL SOCIETY; OF THE CAMBRIDGE PHILOSOPHICAL SOCIETY; OF THE STATISTICAL SOCIETY OF PARIS; OF THE LINNÆAN AND ZOOLOGICAL SOCIETIES; OF THE SOCIETY FOR PROMOTING USEFUL ARTS IN SCOTLAND, ETC.

**WITH ADDITIONS AND NOTES,**

BY JAMES RENWICK, LL.D.

PROFESSOR OF NATURAL EXPERIMENTAL PHILOSOPHY AND CHYMISTRY IN COLUMBIA COLLEGE, NEW YORK.

ILLUSTRATED BY ENGRAVINGS AND WOODCUTS.

---

*SIXTH AMERICAN, FROM THE FIFTH LONDON EDITION, CONSIDERABLY ENLARGED.*

---

NEW YORK:

A. S. BARNES & CO., 51 JOHN-STREET.

CINCINNATI: H. W. DERBY.

1856.

# PREFACE

OF

# THE AMERICAN EDITOR.

---

SEVERAL of the additions which were made by the Editor to the first American edition, have been superseded by the great extension which the original has from time to time received from its author. This is more particularly the case with the sections which had reference to the character of steam at temperatures other than that of boiling water; to the use of steam in navigation; and to its application to locomotion. These sections have of course been omitted. A few new sections and several notes have been added, illustrative of such points as may be most interesting to the American reader.

# PREFACE

TO

# THE FIFTH EDITION.

---

THIS volume should more properly be called a new work than a new edition of the former one. In fact, the book has been almost rewritten. The change which has taken place, even in the short period which has elapsed since the publication of the first edition, in the relation of the steam engine to the useful arts, has been so considerable as to render this inevitable.

The great extension of railroads, and the increasing number of projects which have been brought forward for new lines connecting various points of the kingdom, as well as the extension of steam navigation, not only through the seas and channels surrounding and intersecting these islands, and throughout other parts of Europe, but through the larger waters which are interposed between our dominions in the East and the countries of Egypt and Syria have conferred so much interest on the application of steam to transport, that I have thought it advisable to extend the limits of the present edition considerably beyond those of the last. The chapter on railroads has been enlarged and improved. Three chapters have been added. The twelfth chapter contains a view of steam navigation; the thirteenth contains several important points connected with the economy of steam power, which, when this work was first published, would not have offered sufficient interest to justify their admission into a popular treatise; and the fourteenth chapter contains a series of compendious maxims, for the instruction and guidance of persons desirous of making investments or speculating in railway property.

# PREFACE

TO

## THE FIRST EDITION.

---

THERE are two classes of persons whose attention may be attracted by a treatise on such a subject as the Steam Engine. One consists of those who, by trade or profession, are interested in mechanical science, and who therefore seek information on the subject of which it treats, as a matter of necessity, and a wish to acquire it in a manner and to an extent which may be practically available in their avocations. The other and more numerous class is that part of the public in general, who, impelled by choice rather than necessity, think the interest of the subject itself, and the pleasure derivable from the instances of ingenuity which it unfolds, motives sufficiently strong to induce them to undertake the study of it. Without leaving the former class altogether out of view, it is for the use of the latter principally that the following lectures are designed.

To this class of readers the Steam Engine is a subject which, if properly treated of, must present strong and peculiar attractions. Whether we consider the history of its invention as to time and place, the effects which it has produced, or the means by which it has caused these effects, we find every thing to gratify our national pride, stimulate our curiosity, excite our wonder, and command our admiration. The invention and progressive improvement of this extraordinary machine, is the work of our own time and our own country; it has been produced and brought to perfection almost within the last century, and is the exclusive offspring of British genius fostered and supported by British capital. To enumerate the effects of this invention, would be to count every comfort and luxury of life. It has increased the sum of human happiness, not only by calling new pleasures into existence, but by so cheapening former enjoyments as to render them attainable by those who before never could have hoped to share them. Nor are its effects confined to England alone; they extend over the whole civilized world; and the savage tribes of America, Asia, and Africa, must ere long feel the benefits, remote or immediate, of this all-powerful agent.

If the effect which this machine has had on commerce and the wealth of nations raise our astonishment, the means by which this effect has been produced will not less excite our admiration. The history of the Steam Engine presents a series of contrivances, which, for exquisite and refined ingenuity, stand without a parallel in the annals of human invention. These admirable contrivances, unlike other results of scientific investigation, have also this peculiarity, that to understand and appreciate their excellence requires little previous or subsidiary knowledge. A simple and clear explanation, divested as far as possible of technicalities, and assisted by well selected diagrams, is all that is necessary to render the principles of the construction and operation of the Steam Engine intelligible to a person of a plain understanding and moderate information.

The purpose for which this volume is designed, as already explained, has rendered necessary the omission of many particulars which, however interesting and instructive to the practical mechanic or professional engineer, would have little attraction for the general reader. Our readers require to be informed of the general principles of the construction and operation of Steam Engines, rather than of their practical details. For the same reasons we have confined ourselves to the more striking and important circumstances in the history of the invention and progressive improvement of this machine, excluding many petty disputes which arose from time to time respecting the rights of invention, the interest of which is buried in the graves of their respective claimants.

In the descriptive parts of the work, we have been governed by the same considerations. The application of the force of steam to mechanical purposes has been proposed on various occasions, in various countries, and under a great variety of forms. The list of British patents alone would furnish an author of common industry and application with matter to swell his book to many times the bulk of this volume. By far the greater number of these projects have, however, proved abortive. Descriptions of such unsuccessful, though frequently ingenious machines, we have thought it advisable to exclude from our pages, as not possessing sufficient interest for the readers to whose use this volume is dedicated. We have therefore strictly confined our descriptions either to those Steam Engines which have come into general use, or to those which form an important link in the chain of invention.

*December* 26, 1827.

# CONTENTS.

---

## CHAPTER I.

### PRELIMINARY MATTER.



## CHAPTER II.

### FIRST STEPS IN THE INVENTION.



## CHAPTER III.

### ENGINES OF SAVERY AND NEWCOMEN.

## CHAPTER IV.

ENGINE OF JAMES WATT.



## CHAPTER V.

WATT'S SINGLE-ACTING STEAM ENGINE



## CHAPTER VI.

DOUBLE-ACTING STEAM ENGINE.



## CHAPTER VII.

DOUBLE-ACTING STEAM ENGINE,

(CONTINUED.)



## CHAPTER VIII.

BOILER AND ITS APPENDAGES.

## CHAPTER IX.

### DOUBLE-CYLINDER ENGINES.



## CHAPTER X.

### LOCOMOTIVE ENGINES ON RAILWAYS.



## CHAPTER XI.

### LOCOMOTIVE ENGINES ON TURNPIKE ROADS.



## CHAPTER XII.

### STEAM NAVIGATION.

THE

# STEAM ENGINE

## EXPLAINED AND ILLUSTRATED.

---

## CHAPTER I.

### PRELIMINARY MATTER.

Motion the Agent in Manufactures.—Animal Power.—Power depending on physical Phenomena.—Purpose of a Machine.—Prime Mover.—Mechanical Qualities of the Atmosphere.—Its Weight.—The Barometer.—Fluid Pressure.—Pressure of rarefied Air.—Elasticity of Air.—Bellows.—Effects of Heat.—Thermometer.—Method of making one.—Freezing and boiling Points.—Degrees.—Dilatation of Bodies.—Liquefaction and Solidification.—Vaporization and Condensation.—Latent Heat of Steam.—Expansion of Water in evaporating.—Effects of Repulsion and Cohesion.—Effect of Pressure upon boiling Point.—Formation of a Vacuum by Condensation.

(1.) Of the various productions designed by nature to supply the wants of man, there are few which are suited to his necessities in the state in which the earth spontaneously offers them: if we except atmospheric air, we shall scarcely find another instance: even water, in most cases, requires to be transported from its streams or reservoirs; and food itself, in almost every form, requires culture and preparation. But if, from the mere necessities of physical existence in a primitive state, we rise to the demands of civil and social life—to say nothing of luxuries and refinements—we shall find that every thing which contributes to our convenience, or ministers to our pleasure, requires a previous and extensive expenditure of labour. In most cases, the objects of our enjoyment derive all their excellences, not from any qualities originally inherent in the natural substances out of which they are formed, but from those qualities which have been bestowed upon them by the application of human labour and human skill.

In all those changes to which the raw productions of the earth are submitted in order to adapt them to our wants, one of the principal agents is *motion*. Thus, for example, in the preparation of clothing for our bodies, the various processes necessary for the culture of the cotton require the application of moving power, first to the soil, and subsequently to the plant from which the raw material is obtained: the wool must afterward be picked and cleansed, twisted into threads, and woven into cloth. In all these processes motion is the agent: to cleanse the wool and arrange the fibres of the cotton, the wool must be beaten, teased, carded, and submitted to other processes, by which all the foreign and coarser matter may be separated, and the fibres or threads arranged evenly, side by side. The threads must then receive a rotatory motion, by which they may be twisted into the required form; and finally peculiar motions must be given to them, in order to produce among them that arrangement which characterizes the cloth which it is our final purpose to produce.

In a rude state of society, the motions required in the infant manufactures are communicated by the immediate application of the hand. Observation and reflection, however, soon suggest more easy and effectual means of attaining these ends: the strength of animals is first resorted to for the relief of human labour. Further reflection and inquiry suggest still better expedients. When we look around us in the natural world, we perceive inanimate matter undergoing various effects in which motion plays a conspicuous part: we see the falls of cataracts, the currents of rivers, the eleva tion and depression of the waters of the ocean, the currents of the atmosphere; and the question instantly arises, whether, without sharing our own means of subsistence with the animals whose force we use, we may not equally, or more effectually, derive the powers required from these various phenomena of nature? A difficulty, however, immediately presents itself: we require motion of a particular kind; but wind will not blow, nor water fall as we please, nor as suits our peculiar wants, but according to the fixed laws of nature. We want an *upward* motion; water falls *downward:* we want a *circular* motion; wind blows in a *straight* line. The motions, therefore, which are in actual existence must be modified to suit our purposes: the means whereby these modifications are produced, are called *machines.* A machine, therefore, is an instrument interposed between some natural force or motion, and the object to which force or motion is desired to be transmitted. The construction of the machine is such as to modify the natural motion which is impressed upon it, so that it may transmit to the object to be moved that peculiar species of motion which it is required to have. To give a very obvious example, let us suppose that a circular or rotatory motion is required to be produced, and that the only natural source of motion at our command is a perpendicular fall of water: a wheel is provided, placed upon the axle destined to receive the rotatory motion; this wheel is furnished with cavities in its rim; the water is conducted into the cavities near the top of the wheel on one side; and being caught by these, its weight bears down that side of the wheel, the cavities on the opposite side being empty, and in an inverted position. As the wheel turns, the cavities on the descending side discharge their contents as they arrive near the lowest point, and ascend empty on the other side. Thus a load of water is continually pressing down one side of the wheel, from which the other side is free, and a continued motion of rotation is produced.

In every machine, therefore, there are three objects demanding attention :—first, The power which imparts motion to it, this is called the *prime mover ;* secondly, The nature of the *machine* itself; and thirdly, The object to which the motion is to be conveyed. In the steam engine, the first mover arises from certain phenomena which are exhibited when heat is applied to liquids ; but in the details of the machine and in its application there are several physical effects brought into play, which it is necessary perfectly to understand before the nature of the machine or its mode of operation can be rendered intelligible. We propose, therefore, to devote the present chapter to the explanation and illustration of these phenomena.

(2.) The physical effects most intimately connected with the operations of steam engines are some of the mechanical properties of atmospheric air. The atmosphere is the thin transparent fluid in which we live and move, and which, by respiration, supports animal life. This fluid is apparently so light and attenuated, that it might be at first doubted whether it be really a body at all. It may therefore excite some surprise when we assert, not only that it is a body, but also that it is one of considerable *weight.* We shall be able to prove that it presses on every *square inch** of surface with a weight of about 15lb. avoirdupois.

(3.) Take a glass tube A B (fig. 2) more than 32 inches long, open at one end A, and closed at the other end B, and let it be filled with mercury, (quicksilver.) Let a glass vessel or cistern C, containing a quantity of mercury, be also provided. Applying the finger at A so as to prevent the mercury in the tube from falling out, let the tube be inverted, and the end, stopped by the finger, plunged into the mercury in C. When the end of the tube is below the surface of

* As we shall have frequent occasion to mention this magnitude, it would be well that the reader should be familiar with it. It is a *square*, each side of which is an inch. Such as A B C D, fig. 1.

the mercury in c (fig. 3) let the finger be removed. It will be found that the mercury in the tube will not, as might be expected, fall to the level of the mercury in the cistern c, which it would do were the end B open so as to admit the air into the upper part of the tube. On the other hand, the level D of the mercury in the tube will be about 30 inches above the level c of the mercury in the cistern.

(4.) The cause of this effect is, that the weight of the atmosphere rests on the surface c of the mercury in the cistern, and tends thereby to press it up, or rather to resist its fall in the tube; and as the fall is not assisted by the weight of the atmosphere on the surface D, (since B is closed,) it follows, that as much mercury remains suspended in the tube above the level c as the weight of the atmosphere is able to support.

If we suppose the section of the tube to be equal to the magnitude of a square inch, the weight of the column of mercury in the tube above the level c will be exactly equal to the weight of the atmosphere on each square inch of the surface c. The height of the level D above c being about 30 inches, and a column of mercury two inches in height, and having a base of a square inch, weighing about one pound avoirdupois, it follows that the weight with which the atmosphere presses on each square inch of a level surface is about 15lb. avoirdupois.

An apparatus thus constructed, and furnished with a scale to indicate the height of the level D above the level c, is the *common barometer*. The difference of these levels is subject to a small variation, which indicates a corresponding change in the atmospheric pressure. But we take 30 inches as a standard or average.

(5.) It is an established property of fluids that they press equally in all directions; and air, like every other fluid, participates in this quality. Hence it follows, that since the downward pressure or weight of the atmosphere is about 15lb. on the square inch, the lateral, upward, and oblique pressures are of the same amount. But, independently of the general principle, it may be satisfactory to give experimental proof of this.

Let four glass tubes A, B, C, D, (fig. 4,) be constructed of sufficient length, closed at one end end A, B, C, D, and open at the other. Let the open ends of three of them be bent, as represented in the tubes B, C, D. Being previously filled with mercury, let them all be gently inverted so as to have their closed ends up as here represented. It will be found that the mercury will be sustained in all,* and that the difference of the levels in all will be the same. Thus the mercury is sustained in A by the upward pressure of the atmosphere, in B by its horizontal or lateral pressure, in C by its downward pressure, and in D by its oblique pressure; and as the difference of the levels is the same in all, these pressures are exactly equal.

(6.) In the experiment described in (3), the space B D (fig. 3) at the top of the tube from which the mercury has fallen, is perfectly void and empty, containing neither air nor any other fluid: it is called therefore a *vacuum*. If, however, a small quantity of air be introduced into that space, it will immediately begin to exert a pressure on D, which will cause the surface D to descend, and it will continue to descend until the column of mercury C D is so far diminished that the weight of the atmosphere is sufficient to sustain it, as well as the pressure exerted upon it by the air in the space B D.

The quantity of mercury which falls from the tube in this case is necessarily an equivalent for the pressure of the air introduced, so that the pressure of this air may be exactly ascertained by allowing about one pound per square inch for every two inches of mercury which has fallen from the tube. The pressure of the air or any other fluid above the

* This experiment with the tube A requires to be very carefully executed, and the tube should be one of small bore.

mercury in the tube, may at once be ascertained by comparing the height of the mercury in the tube with the height of the barometer; the difference of the heights will always determine the pressure on the surface of the mercury in the tube. This principle will be found of some importance in considering the action of the modern steam engines.

The air which we have supposed to be introduced into the upper part of the tube, presses on the surface of the mercury with a force much greater than its weight. For example, if the space B D (fig. 3) were filled with atmospheric air in its ordinary state, it would exert a pressure on the surface D equal to the whole pressure of the atmosphere, although its weight might not amount to a single grain. The property in virtue of which the air exerts this pressure is its *elasticity*, and this force is diminished in precisely the proportion in which the space which the air occupies is increased.

Thus it is known that atmospheric air in its ordinary state exerts a pressure on the surface of any vessel in which it is confined, amounting to about 15lb. on every square inch. If the capacity of the vessel which contains it be doubled, it immediately expands and fills the double space, but in doing so it loses half its elastic force, and presses only with the force of 7½lb. on every square inch. If the capacity of the vessel had been enlarged five times, the air would still have expanded so as to fill it, but would exert only a fifth part of its first pressure, or 3lb. on every square inch.

This property of losing its elastic force as its volume or bulk is increased, is not peculiar to air. It is common to all elastic fluids, and we accordingly find it in steam; and it is absolutely necessary to take account of it in estimating the effects of that agent.

(7.) There are numerous instances of the effects of these properties of atmospheric air which continually fall under our observation. If the nozzle and valve-hole of a pair of bellows be stopped, it will require a very considerable force to separate the boards. This effect is produced by the diminished elastic force of the air remaining between the boards upon the least increase of the space within the bellows, while the atmosphere presses, with undiminished force, on the external surfaces of the boards. If the boards be separated so as to double the space within, the elastic force of the included air will be about 7½lb. on every square inch, while the pressure on the external surfaces will be 15lb. on every square inch; consequently, it will require as great a force to sustain the boards in such a position, as it would to separate them if each board were forced against the other with a pressure of 7½lb. per square inch on their external surfaces.

When boys apply a piece of moistened leather to a stone, so as to exclude the air from between them, the stone, though it be of considerable weight, may be lifted by a string attached to the leather: the cause of which is the atmospheric pressure, which keeps the leather and the stone in close contact.

(8.) The next class of physical effects which it is necessary to explain, are those which are produced when heat is imparted or abstracted from bodies.

In general, when heat is imparted to a body, an enlargement of bulk will be the immediate consequence, and at the same time the body will become warmer to the touch. These two effects of expansion and increase of warmth going on always together, the one has been taken as a measure of the other; and upon this principle the common thermometer is constructed. That instrument consists of a tube of glass, terminated in a bulb, the magnitude of which is considerable, compared with the bore of the tube. The bulb and part of the tube are filled with mercury, or some other liquid. When the bulb is exposed to any source of heat, the mercury contained in it, being warmed or increased in temperature, is at the same time increased in bulk, or expanded or dilated, as it is called. The bulb not having sufficient capacity to contain the increased bulk of mercury, the liquid is forced up in the tube, and the quantity of expansion is determined by observing the ascent of the column in the tube.

An instrument of this kind, exposed to heat or cold, will fluctuate accordingly, the mercury rising as the heat to which it is exposed is increased, and falling by exposure to cold. In order, however, to render it an accurate measure of temperature, it is necessary to connect with it a scale by which the elevation or depression of the mercury in the tube may be measured. Such a scale is constructed for thermometers in this country in the following manner:—Let us suppose the instrument immersed in a vessel of melting ice: the column of mercury in the tube will be observed to fall to a certain point, and there maintain its position unaltered: let that point be marked upon the tube. Let the instrument be now transferred to a vessel of boiling water at a time when the barometer stands at the altitude of 30 inches: the mercury in the tube will be observed to rise until it attain a certain elevation, and will there maintain its position. It will be found, that though the water continue to be exposed to the action of the fire, and continue to boil, the mercury in the tube will not continue to rise, but will maintain a fixed position: let the point to which the mercury has risen, in this case, be likewise marked upon the tube.

The two points, thus determined, are called the *freezing* and the *boiling* points. If the distance upon the tube between these two points be divided into 180 equal parts, each of these parts is called a *degree;* and if this division be continued, by taking equal divisions below the freezing point, until 32 divisions be taken, the last division is called the *zero*, or *naught* of the thermometer. It is the point to which the mercury would fall, if the thermometer were immersed in a certain mixture of snow and salt. When thermometers were first invented, this point was taken as the zero point, from an erroneous supposition that the temperature of such a mixture was the lowest possible temperature.

The degrees upon the instrument thus divided are counted

upward from the zero, and are expressed, like the degrees of a circle, by placing a small ° over the number. Thus it will be perceived that the freezing point is 32° of our thermometer, and the boiling point will be found by adding 180° to 32°; it is therefore 212°.

The temperature of a body is that elevation to which the thermometer would rise when the mercury enclosed in it would acquire the same temperature. Thus, if we should immerse the thermometer, and should find that the mercury would rise to the division marked 100°, we should then affirm that the temperature of the water was 100°.

(9.) The dilatation which attends an increase of temperature is one of the most universal effects of heat. It varies, however, in different bodies: it is least in solid bodies; greater in liquids; and greatest of all in bodies in the aeriform state. Again, different solids are differently susceptible of this expansion. Metals are the most susceptible of it; but metals of different kinds are differently expansible.

As an increase of temperature causes an increase of bulk, so a diminution of temperature causes a corresponding diminution of bulk, and the same body always has the same bulk at the same temperature.

A flaccid bladder, containing a small quantity of air, will, when heated, become quite distended; but it will again resume its flaccid appearance when cold. A corked bottle of fermented liquor, placed before the fire, will burst by the effort of the air contained in it to expand when heated.

Let the tube A B (fig. 5) open at both ends, have one end inserted in the neck of a vessel C D, containing a coloured liquid, with common air above it; and let the tube be fixed so as to be air-tight in the neck: upon heating the vessel, the warm air enclosed in the vessel C D above the liquid will begin to expand, and will press upon the surface of the liquid, so as to force it up in the tube A B.

In bridges and other structures, formed of iron, mechanical provisions are introduced to prevent the fracture or strain which would take place by the expansion and contraction which the metal must undergo by the changes of temperature at different seasons of the year, and even at different hours of the day.

Thus all nature, animate and inanimate, organized and unorganized, may be considered to be incessantly breathing heat ; at one moment drawing in that principle through all its dimensions, and at another moment dismissing it.

(10.) Change of bulk, however, is not the only nor the most striking effect which attends the increase or diminution of the quantity of heat in a body. In some cases, a total change of form and of mechanical qualities is effected by it. If heat be imparted in sufficient quantity to a solid body, that body, after a certain time, will be converted into a liquid. And again, if heat be imparted in sufficient quantity to this liquid, it will cease to exist in the liquid state, and pass into the form of vapour.

By the abstraction of heat, a series of changes will be produced in the opposite order. If from the vapour produced in this case, a sufficient quantity of heat be taken, it will return to the liquid state ; and if again from this liquid heat be further abstracted, it will at length resume its original solid state.

The transmission of a body from the solid to the liquid state, by the application of heat, is called *fusion* or *liquefaction;* and the body is said to be fused, *liquefied*, or melted.

The reciprocal transmission from the liquid to the solid state, is called *congelation* or *solidification ;* and the liquid is said to be *congealed* or *solidified.*

The transmission of a body from the liquid to the vaporous or aeriform state is called *vaporization;* and the liquid is said to be *vaporized* or *evaporated.*

The reciprocal transmission of vapour to the liquid state is called *condensation;* and the vapour is said to be *condensed.*

We shall now examine more minutely the circumstances which attend these remarkable and important changes in the state of body.

(11.) Let us suppose that a thermometer is imbedded in any solid body; for example, in a mass of sulphur; and that it stands at the ordinary temperature of 60 degrees: let the sulphur be placed in a vessel, and exposed to the action of fire. The thermometer will now be observed gradually to rise, and it will continue to rise until it exhibit the temperature of 218°. Here, however, notwithstanding the continued action of the fire upon the sulphur, the thermometer will become stationary; proving, that notwithstanding the supply of heat received from the fire, the sulphur has ceased to become hotter. At the moment that the thermometer attains this stationary point, it will be observed that the sulphur has commenced the process of fusion; and this process will be continued, the thermometer being stationary, until the whole mass has been liquefied. The moment the liquefaction is complete, the thermometer will be observed again to rise, and it will continue to rise until it attain the elevation of 570°. Here, however, it will once more become stationary; and notwithstanding the heat supplied to the sulphur by the fire, the liquid will cease to become hotter: when this happens, the sulphur will boil; and if it continue to be exposed to the fire a sufficient length of time, it will be found that its quantity will gradually diminish, until at length it will all disappear from the vessel which contained it. The sulphur will, in fact, be converted into vapour.

From this process we infer, that all the heat supplied during the processes of liquefaction and vaporization is consumed in effecting these changes in the state of the body; and that, under such circumstances, it does not increase the temperature of the body on which the change is produced.

These effects are general: all solid bodies would pass into the liquid state by a sufficient application of heat; and all liquid bodies would pass into the vaporous state by the same means. In all cases the thermometer would be stationary during these changes, and consequently the temperature of the body, in those periods, would be maintained unaltered.

(12.) Solids differ from one another in the temperatures at which they become liquid. These temperatures are called their *melting points*. Thus the melting point of ice is 32°; that of lead 612°; that of gold 5237°.* The heat which is supplied to a body during the processes of fusion or vaporization, and which does not affect the thermometer, or increase the temperature of the body fused or vaporized, is said to become *latent*. It can be proved to exist in the body fused or vaporized, and may even be taken from that body. In parting with it the body does not fall in temperature, and consequently the loss of this heat is not indicated by the thermometer any more than its reception. The term latent heat is merely intended to express this fact, of the thermometer being insensible to the presence or absence of this portion of heat, and is not intended to express any theoretical notions concerning it.

(13.) In explaining the construction and operation of the steam engine, although it is necessary occasionally to refer to the effects of heat upon bodies in general, yet the body, which is by far the most important to be attended to, so far as the effects of heat upon it are concerned, is water. This body is observed to exist in the three different states, the solid, the liquid, and the vaporous, according to the varying temperature to which it is exposed. All the circumstances which have been explained in reference to metals, and the substance sulphur in particular, will, *mutatis mutandis*, be applicable to water. But in order perfectly to comprehend the properties of the steam engine, it is necessary to

* Temperatures above 650° cannot be measured by the mercurial thermometer. They can be inferred onlv with probability by pyrometers.

render a more rigorous and exact account of these phenomena, so far as they apply to the changes produced upon water by the effects of heat.

Let us suppose a mass of ice immersed in the mixture of snow and salt which determines the zero point of the thermometer; this mass, if allowed to continue a sufficient length of time submerged in the mixture, will necessarily acquire its temperature, and the thermometer immersed in it will stand at zero. Let the ice be now withdrawn from the mixture, still keeping the thermometer immersed in it, and let it be exposed to the atmosphere at the ordinary temperature, say 60°. At first the thermometer will be observed gradually and continuously to rise until it attain the elevation of 32°; it will then become stationary, and the ice will begin to melt: the thermometer will continue standing at 32° until the ice shall be completely liquefied. The liquid ice and the thermometer being contained in the same vessel, it will be found, when the liquefaction is completed, that the thermometer will again begin to rise, and will continue to rise until it attain the temperature of the atmosphere, viz. 60°. Hitherto the ice or water has received a supply of heat from the surrounding air; but now an equilibrium of temperature having been established, no further supply of heat can be received; and if we would investigate the further effects of increased heat, it will be necessary to expose the liquid to fire, or some other source of heat. But previous to this, let us observe the time which the thermometer remains stationary during the liquefaction of the ice: if noted by a chronometer, it would be found to be a hundred and forty times the time during which the water in the liquid state was elevated one degree; the inference from which is, that in order to convert the solid ice into liquid water, it was necessary to receive from the surrounding atmosphere one hundred and forty times as much heat as would elevate the liquid water one degree in temperature; or, in other words, that to liquefy a given weight of ice requires as much heat as would raise the same weight of water 140° in temperature: or from 32° to 172°.

The latent heat of water acquired in liquefaction is there fore 140°.

(14.) Let us now suppose that, a spirit lamp being applied to the water already raised to 60°, the effects of a further supply of heat be observed: the thermometer will continue to rise until it attain the elevation of 212°, the barometer being supposed to stand at 30 inches. The thermometer having attained this elevation will cease to rise; the water will therefore cease to become hotter, and at the same time bubbles of steam will be observed to be formed at the bottom of the vessel containing the water, near the flame of the spirit lamp. These bubbles will rise through the water, and escape at the surface, exhibiting the phenomena of ebullition, and the water will undergo the process of *boiling*.

During this process, the thermometer will constantly be maintained at the same elevation of 212°; but if the time be noted, it will be found that the water will be altogether evaporated, if the same source of heat be continued to be applied to it six and a half times as long as was necessary to raise it from the freezing to the boiling point. Thus, if the application of the lamp to water at 32°, be capable of raising that water to 212° in one hour, the same lamp will require to be applied to the boiling water for six hours and a half, in order to convert the whole of it into steam. Now if the steam into which it is thus converted were carefully preserved in a receiver, maintained at the temperature of 212°, this steam would be found to have that temperature, and not a greater one; but it would be found to fill a space about 1700 times greater than the space it occupied in the liquid state, and it would possess an elastic force equal to the pressure of the atmosphere under which it was boiled; that is to say, it would press the sides of the vessel which contained it with a pressure equivalent to that of a column of mercury of 30 inches in height; or what is the same thing, at the rate of about 15lb. on every square inch of surface.

(15.) As the quantity of heat expended in raising the water from 32° to 212°, is 180°; and as the quantity of heat necessary to convert the same water into steam is six and a half times this quantity, it follows, that the quantity of heat requisite for converting a given weight of water into steam, will be found by multiplying 180° by 5½. The product of these numbers being 990°, it follows, that, to convert a given weight of water at 212° into steam of the same temperature, under the pressure of the atmosphere, when the barometer stands at 30 inches, requires as much heat as would be necessary to raise the same water 990° higher in temperature. The heat, not being sensible to the thermometer, is latent heat; and accordingly it may be stated, that the latent heat, necessary to convert water into steam under this pressure is, in round numbers, 1000°.

(16.) All the effects of heat which we have just described may be satisfactorily accounted for, by supposing that the principle of heat imparts to the constituent atoms of bodies a force, by virtue of which they acquire a tendency to repel each other. But in conjunction with this, it is necessary to notice another force, which is known to exist in nature: there is observable among the corpuscles of bodies a force, in virtue of which they have a tendency to cohere, and collect themselves together in solid concrete masses: this force is called the attraction of cohesion. These two forces—the natural cohesion of the particles, and the repulsive energy introduced by heat—are directly opposed to one another, and the state of the body will be decided by the predominance of the one or the other, or their mutual equilibrium. If the natural cohesion of the constituent particles of the body considerably predominate over the repulsive energy introduced by the heat, then the cohesion will take effect; the particles of the body will coalesce, the mass will become rigid and solid, and the particles will hold together in one invariable mass, so that they cannot drop asunder by the mere effect of their weight. In such cases, a more or less considerable force must be applied, in order to break the body, or to tear its parts asunder. Such is the quality which characterizes the state which, in mechanics, is called the state of solidity.

If the repulsive energy introduced by the application of heat be equal, or nearly equal, to the natural cohesion with which the particles of the body are endued, then the predominance of the cohesive force may be insufficient to resist the tendency which the particles may have to drop asunder by their weight. In such a case, the constituent particles of the body cannot cohere in a solid mass, but will separate by their weight, fall asunder, and drop into the various corners, and adapt themselves to the shape of any vessel in which the body may be contained. In fact, the body will take the liquid form. In this state, however, it does not follow that the cohesive principle will be altogether inoperative: it may, and does, in some cases, exist in a perceptible degree, though insufficient to resist the separate gravitation of the particles. The tendency which particles of liquids have, in some cases, to collect in globules, plainly indicates the predominance of the cohesive principle: drops of water collected upon the window pane; drops of rain condensed in the atmosphere; the tear which trickles on the cheek; drops of mercury, which glide over any flat surface, and which it is difficult to subdivide or scatter into smaller parts; are all obvious indications of the predominance of the cohesive principle in liquids.

By the due application of heat, even this small degree of cohesion may be conquered, and a preponderance of the opposite principle of repulsion may be created. But another physical influence here interposes its aid, and conspires with cohesion in resisting the transmission of the body from the liquid to the vaporous state: this force is no other than the pressure of the atmosphere, already explained. This pressure has an obvious tendency to restrain the particles of the liquid, to press them together, and to resist their separation. The repulsive principle of the heat introduced must therefore not only neutralize the cohesion, but must also impart to the atoms of the liquid a sufficient elasticity or repulsive energy to enable them to fly asunder, and assume the vaporous form in spite of this atmospheric resistance.

Now, it is clear that if this atmospheric resistance be subject to any variation in its intensity, from causes whether natural or artificial, the repulsive energy necessary to be introduced by the heat will vary proportionally: if the atmospheric pressure be diminished, then less heat will be necessary to vaporize the liquid. If, on the other hand, this pressure be increased, a greater quantity of heat will be required to impart the necessary elasticity.

(17.) From this reasoning we must expect that any cause, whether natural or artificial, which diminishes the atmospheric pressure upon the surface of a liquid, will cause that liquid to boil at a lower temperature: and on the other hand, any cause which may increase the atmospheric pressure upon the liquid, will render it necessary to raise it to a higher temperature before it can boil.

These inferences we accordingly find supported by experience. Under a pressure of 15lb. on the square inch, *i. e.* when the barometer is at 30 inches, water boils at the temperature of 212° of the common thermometer. But if water at a lower temperature, suppose 180°, be placed under the receiver of an air-pump, and, by the process of exhaustion, the atmospheric pressure be removed, or very much diminished, the water will boil, although its temperature still remain at 180°, as may be indicated by a thermometer placed in it.

On the other hand, if a thermometer be inserted air-tight in the lid of a close digester containing water with common atmospheric air above it, when the vessel is heated the air acquires an increased elasticity; and being confined by the cover, presses, with increased force, on the surface of the water. By observing the thermometer while the vessel is exposed to the action of heat, it will be seen to rise considerably above 212°, suppose to 230°, and would continue so to rise until the strength of the vessel could no longer resist the pressure within it.

The temperature at which water boils is commonly said to be 212°, which is called *the boiling point* of the thermometer; but, strictly speaking, this is only true when the barometer stands at 30 inches. If it be lower, water will boil at a lower temperature, because the atmospheric pressure is less; and if it be higher, as at 31, water will not boil until it receives a higher temperature, the pressure being greater.

According as the cohesive forces of the particles of liquids are more or less active, they boil at greater or less temperatures. In general the lighter liquids, such as *alcohol* and *ether*, boil at lower temperatures. These fluids, in fact, would boil by merely removing the atmospheric pressure, as may be proved by placing them under the receiver of an air-pump, and withdrawing the air. From this we may conclude that these and similar substances would never exist in the liquid state at all, but for the atmospheric pressure.

(18.) The elasticity of vapour raised from a boiling liquid, is equal to the pressure under which it is produced. Thus, steam raised from water at 212°, and, therefore, under a pressure of 15lb. on the square inch, is endued with an elastic force which would exert a pressure on the sides of any vessel which confines it, also equal to 15lb. on the square inch. Since an increased pressure infers an increased temperature in boiling, it follows, that where steam of a higher pressure than the atmosphere is required, it is necessary that the water should be boiled at a higher temperature.

(19.) We have already stated that there is a certain point at which the temperature of a liquid will cease to rise, and that all the heat communicated to it beyond this is consumed in the formation of vapour. It has been ascertained, that when water boils at 212°, under a pressure equal to 30 inches of mercury, a cubic inch of water forms a cubic foot* of steam, the elastic force of which is equal to the atmospheric pressure, and the temperature of which is 212°. Since there are 1728 cubic inches in a cubic foot, it follows, that when water at this temperature passes from the liquid to the vaporous state, it is dilated into 1728 times its bulk.

(20.) We have seen that about 1000 degrees of heat must be communicated to any given quantity of water at 212°, in order to convert it into steam of the same temperature, and possessing a pressure amounting to about 15 pounds on the square inch, and that such steam will occupy above 1700 times the bulk of the water from which it was raised. Now we might anticipate, that by abstracting the heat thus employed in converting the liquid into vapour, a series of changes would be produced exactly the reverse of those already described; and such is found to be actually the case. Let us suppose a vessel, the capacity of which is 1728 cubic inches, to be filled with steam, of the temperature of 212°, and exerting a pressure of 15 pounds on the square inch; let 5½ cubic inches of water, at the temperature of 32°, be injected into this vessel, immediately the steam will impart the heat, which it has absorbed in the process of vaporization, to the water thus injected, and will itself resume the liquid form. It will shrink into its primitive dimensions of one cubic inch, and the heat which it will dismiss will be sufficient to raise the 5½ cubic inches of injected water to the temperature of 212°. The contents of

* The terms *cubic inch* and *cubic foot* are easily explained. A common die, used in games of chance, has the figure which is called a cube. It is a solid having twelve straight edges equal to one another. It has six sides, each of which is square, and which are also equal to one another. If its edges be each one inch in length, it is called a *cubic inch;* if one foot, a *cubic foot;* if one yard, a *cubic yard,* &c. This figure is represented in perspective, in fig. 6.

the vessel will thus be 6½ cubic inches of water at the temperature of 212°. One of these cubic inches is in fact the steam which previously filled the vessel reconverted into water, the other 5½ are the injected water which has been raised from the temperature of 32° to 212° by the heat which has been dismissed by the steam in resuming the liquid state. It will be observed that in this transmission no temperature is lost, since the cubic inch of water into which the steam is converted has the same temperature as the steam had before the cold water was injected.

These consequences are in perfect accordance with the results already obtained from observing the time necessary to convert a given quantity of water into steam by the application of heat. From the present result it follows, that in the reduction of a given quantity of steam to water it parts with as much heat as was sufficient to raise 5½ cubic inches from 32° to 212°, that is, 5½ times 180° or 990°.

(21.) There is an effect produced in this process to which it is material that we should attend. The steam which filled the space of 1728 cubic inches shrinks when reconverted into water into the dimensions of 1 cubic inch. It therefore leaves 1727 cubic inches of the vessel it contains unoccupied. By this property steam is rendered instrumental in the formation of a vacuum.

By allowing steam to circulate through a vessel, the air may be expelled from it, and its place filled by steam. If the vessel be then closed and cooled, the steam will be reduced to water, and, falling in drops on the bottom and sides of the vessel, the space which it filled will become a *vacuum*. This may be easily established by experiment. Let a long glass tube be provided with a hollow ball at one end, and having the other end open.* Let a small quantity of spirits be poured in at the open end, and placing the glass ball over the flame of a lamp, let the spirits be boiled. After

* A common glass flask with a long neck will answer the purpose

D

some time the steam will be observed to issue copiously from the open end of the tube which is presented upward. When this takes place, let the tube be inverted, and its open end plunged in a basin of cold water. The heat being thus removed, the cool air will reconvert the steam in the tube into liquid, and a vacuum will be produced, into which the pressure of the atmosphere on the surface of the water in the basin will force the water through the tube, and it will rush up with considerable force, and fill the glass ball.

In this experiment it is better to use spirits than water, because they boil at a lower heat, and expose the glass to less liability to break, and also the tube may more easily be handled.

---

## CHAPTER II.

### FIRST STEPS IN THE INVENTION.

Futility of early claims.—Watt, the real inventor.—Hero of Alexandria.—Blasco Garay.—Solomon de Caus.—Giovanni Branca.—Marquis of Worcester.—Sir Samuel Morland.—Denis Papin.—Thomas Savery.

(22.) In the history of the progress of the useful arts and manufactures, there is perhaps no example of any invention the credit of which has been so keenly contested as that of the steam engine. Claims to it have been advanced by different nations, and by different individuals of the same nation. The partisans of the competitors for this honour have argued their pretensions, and pressed their claims, with a zeal which has occasionally outstripped the bounds of discretion; and the contest has not unfrequently been tinged with prejudices, both national and personal, and marked with a degree of asperity quite unworthy of so noble a cause, and altogether beneath the dignity of science.

The efficacy of the steam engine considered as a mechanical agent depends, first, on the several physical properties from which it derives its operation, and, secondly, on the various pieces of mechanism and details of mechanical arrangement by which these properties are rendered practically available. If the merit of the invention must be ascribed to the discoverer and contriver of these, then the contest will be easily decided, because it will be obvious that the prize is not due to any one individual, but must be distributed in different proportions among several. If, however, he is best entitled to the credit of the invention, who has by the powers of his mechanical genius imparted to the machine that form and those qualities from which it has received its present extensive utility, and by which it has become an agent of transcendent power, which has spread its beneficial effects throughout every part of the civilized globe, then the universal consent of mankind will, as it were by acclamation, award the prize to one individual, whose pre-eminent genius places him far above all other competitors, and from the applicatiun of whose mental energies to this machine may be dated those grand effects which have rendered it a topic of interest to every individual for whom the progress of human civilization has any attractions. Before the era marked by the discoveries of JAMES WATT, the steam engine, which has since become an object of such universal interest, was a machine of extremely limited power, greatly inferior in importance to most other mechanical contrivances used as prime movers. But from that time it is scarcely necessary here to state that it became a subject not of British interest only, but one with which the progress of the human race became intimately mixed up

Since, however, the question of the progressive developement of those physical principles on which the steam engine depends, and of their mechanical application, has of late years received some importance, as well from the interest which the public manifest toward them as from the rank of the writers who have investigated them, we have thought it expedient to state briefly, but we trust with candour and fairness, the successive steps which appear to have led to this invention.

The engine as it exists at present is not, strictly speaking, the exclusive invention of any one individual: it is the result of a series of discoveries and inventions which have for the last two centuries been accumulating. When we attempt to trace back its history, and to determine its first inventor, we experience the same difficulty as is felt in tracing the head of a great river: as we ascend its course, we are embarrassed by the variety of its tributary streams, and find it impossible to decide which of those channels into which it ramifies ought to be regarded as the principal stream; and it terminates at length in a number of threads of water, each in itself so insignificant as to be unworthy of being regarded as the source of the majestic object which has excited the inquiry.

From a very early period the effects of heat upon liquids, and more especially the production of steam or vapour, was regarded as a probable source of mechanical power, and numerous speculators directed their attention to it, and exerted their inventive faculties to derive from it an effective mover. It was not, however, until the commencement of the eighteenth century that any invention was produced which was practically applied, even unsuccessfully. All the attempts previous to that time were either suggestions which were limited to paper or experiments confined to models; or, if they exceeded this, they never outlived a single trial on a larger scale. Nevertheless many of these suggestions and experiments, being recorded and accessible to future inquirers, doubtless offered useful hints and some practical aid to those more successful investigators who subsequently contrived engines in such forms as to be practically available on a large scale for mechanical purposes. It is right and just, therefore—mere suggestions and abortive experiments though they may have been—to record them, that each inventor and discoverer may receive the just credit due to his share in this splendid mechanical invention. We shall then in the present chapter briefly enumerate, in chronological order, the successive steps so far as they have come to our knowledge.

### HERO OF ALEXANDRIA, 120 B C.

(23.) In a work entitled *Spiritalia seu Pneumatica*, one of the numerous works of this philosopher which has remained to us, is contained a description of a machine moved by vapour of water. A hollow sphere, of which A B represents a section, is supported on two pivots at A and B, which are the extremities of tubes A C D and B E F, which pass into a boiler where steam is generated. This steam flows through small apertures at the extremities A and B, and fills the hollow sphere. One or more horizontal arms K G, I H, project from this sphere, and are likewise filled with steam, but are closed at their extremities. Conceive a

*Fig. 1.*

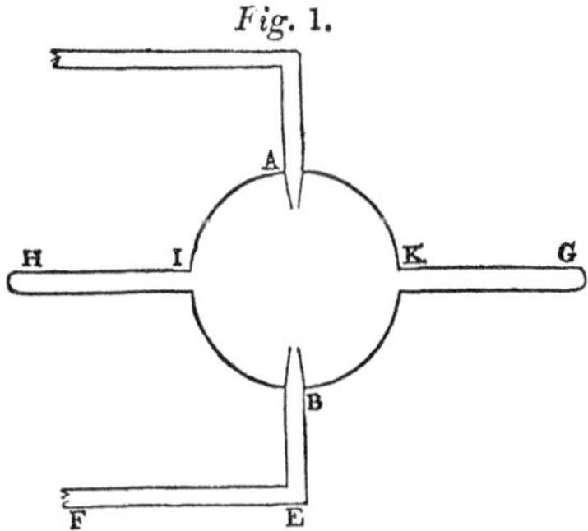


small hole made near the extremity G, but at one side of one of the tubes; the steam confined in the tube and globe would immediately rush from the hole with a force proportionate to its pressure within the globe. On the common principle of mechanics a reaction would be produced, and the tube would recoil in the same manner as a gun when discharged. The tubular arm K G being thus pressed in a direction opposed to that in which the steam issues, the sphere would revolve accordingly, and would continue to revolve so long as the steam would continue to flow from the aperture. The force of recoil would be increased by making a similar aperture in two or more arms, care being taken that all the apertures should be placed so as to cause the sphere to revolve in the same direction.

This motion being once produced might be transmitted by ordinary mechanical contrivance to any machinery which its power might be adequate to move.

This method of using steam is not adopted in any part or any form of the modern steam engine.

### BLASCO DE GARAY, A. D. 1543.

(24.) In the year 1826 there appeared in Zach's Correspondence a communication from Thomas Gonsalez, Director of the Royal Archives of Simancas, giving an account of an experiment reported to have been made in the year 1543 by order of Charles V. in the port of Barcelona. Blasco de Garay, a sea captain, had contrived a machine by which he proposed to propel vessels without oars or sails. Garay concealed altogether the nature of the machine which he used: all that was seen during the experiment was, that it consisted of a great boiler for water, and that wheels were kept in revolution at each side of the vessel. The experiment was made upon a vessel called the Trinity, of 200 tons burden, and was witnessed by several official personages, whose presence on the occasion was commanded by the king. One of the witnesses reported that it was capable of moving the vessel at the rate of two leagues in three hours, that the machine was too complicated and expensive, and was exposed to the danger of explosion. The other witnesses, however, reported more favourably. The result of the experiment was thought to be favourable: the inventor was promoted, and received a pecuniary reward, besides having all his expenses defrayed.

From the circumstance of the nature of the impelling power having been concealed by the inventor, it is impossible to say in what this machine consisted, or even whether steam exerted any agency whatever in it, or, if it did, whether it might not have been, as was most probably the case, a reproduction of Hero's contrivance. It is rather unfavourable to the claims advanced by the advocates of the Spaniard, that although it is admitted that he was rewarded and promoted in consequence of the experiment, yet it does not appear that it was again tried, much less brought into practical use.

### SOLOMON DE CAUS, 1615.

(25.) A work entitled "Les Raisons des Forces Mouvantes, avec diverses Machines tant utiles que plaisantes," published at Frankfort in 1615, by Solomon de Caus, a native of France, contains the following theorem:—

"*Water will mount by the help of fire higher than its level,*" which is explained and proved in the following terms:

"The third method of raising water is by the aid of fire. On this principle may be constructed various machines: I shall here describe one. Let a ball of copper marked A, well soldered in every part, to which is attached a tube and stopcock marked D, by which water may be introduced; and also another tube marked B C, which will be soldered into the top of the ball, and the lower end C of which shall descend nearly to the bottom of the ball without touching it. Let the said ball be

*Fig. 2.*

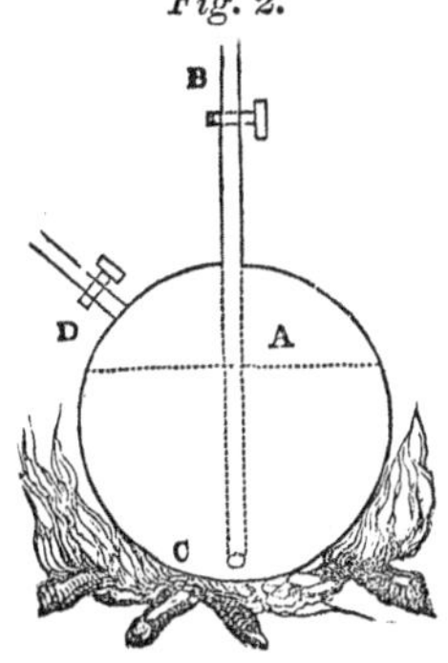

filled with water through the tube D, then shutting the stop-cock D, and opening the stop-cock in the vertical tube B C, let the ball be placed upon a fire, the heat acting upon the said ball will cause the water to rise in the tube B C."

Such is the description of the apparatus of De Caus as given by himself; and on this has been founded a claim to the invention of the steam engine. It will be observed, that neither in the original theorem, nor in the description of the machine which accompanies it, is the word *steam* anywhere used. Now it was well known, by all conversant in physics, long before the date of the publication containing this description, that atmospheric air when heated acquires an increased elastic force. As the experiment is described, the other part of the ball A is filled with atmospheric air; the heat of the fire acting upon the air through the external surface of the ball, and likewise transmitted through the water, would of course raise the temperature of the air contained in the vessel, would thereby increase its elasticity, and would cause the water to rise in the tube B C, upon a physical principle altogether independent of the qualities of steam. The effect produced, therefore, is just what might have been expected by any one acquainted with the common properties of air, though entirely ignorant of those of steam; and in point of fact, the pressure of the air is as much concerned in this case in raising the water as the pressure of the steam.

This objection, however, is combatted by another theorem contained in the same work, in which De Caus speaks of "the strength of the vapour produced by the action of the fire, which causes water to mount; which vapour will issue from the stop-cock with great violence after the water has been expelled."

If De Caus be admitted to have understood the elastic property of the vapour of water, and to have attributed the ascent of the water in the tube C B to the pressure of that vapour upon the surface of the water confined in the copper ball, it must be admitted that he suggested one of the ways of using the power of steam as a mechanical agent. In the modern steam engine this pressure is not now used against a liquid surface, but against the solid surface of a piston. This, however, should not take from De Caus whatever credit be due to the suggestion of the physical property in question.

### GIOVANNI BRANCA, 1629.

(26.) In a work published at Rome in 1629, entitled "Le Machine del G. Branca," is contained a description of a machine for propelling a wheel by a blast of steam. This contrivance consists of a wheel furnished with flat vanes upon its rim, like the boards of a paddle wheel. The steam is produced in a close vessel, and made to issue with violence from the extremity of a pipe. Being directed against the vanes, it causes the wheel to revolve, and this motion may be imparted by the usual mechanical contrivances to any machinery which it was intended to move.

This contrivance has no analogy whatever to any part of the modern steam engines in any of their various forms.

### EDWARD SOMERSET, MARQUIS OF WORCESTER, 1663.

(27.) Of all the individuals to whom the invention of the steam engine has been ascribed, the most celebrated was the Marquis of Worcester, the author of a work entitled "The Scantling of One Hundred Inventions," but which is more commonly known by the title "A Century of Inventions." It is to him that by far the greater number of writers and inquirers on this subject ascribe the merit of the discovery of the invention. This contrivance is described in the following terms in the sixty-eighth invention in the work above named:—

"I have invented an admirable and forcible way to drive up water by fire; not by drawing or sucking it upwards, for that must be, as the philosopher terms it, *infra spæram activitatis*, which is but at such a distance. But this way hath no bounder if the vessels be strong enough. For I have taken a piece of whole cannon whereof the end was burst, and filled it three quarters full of water, stopping and screwing up the broken end, as also the touch-hole, and making a constant fire under it; within twenty-four hours, it burst and made a great crack. So that, having a way to make my vessels so that they are strengthened by the force within them, and the one to fill after the other, I have seen the water run like a constant fountain stream forty feet high. One vessel of water rarefied by fire driveth up forty of cold water, and a man that tends the work has but to turn two cocks; that one vessel of water being consumed, another begins to force and refill with cold water, and so successively; the fire being tended and kept constant, which the selfsame person may likewise abundantly perform in the interim between the necessity of turning the said cocks."

These experiments must have been made before the year 1663, in which the "Century of Inventions" was published. The description of the machine here given, like other descriptions in the same work, was only intended to express the effects produced, and the physical principle on which their production depends. It is, however, sufficiently explicit to enable any one conversant with the subsequent contrivance of Savery, to perceive that Lord Worcester must have contrived a machine containing all that part of Savery's engine in which the direct force of steam is employed. As in the above description, the separate boiler or generator of steam is distinctly mentioned; that the steam from this is conducted into another vessel containing the cold water to be raised; that this water is raised by the pressure of steam acting upon its surface; that when one vessel of water has thus been discharged, the steam acts upon the water contained in another vessel, while the first is being replenished; and that a continued upward current of water is maintained by causing the steam to act alternately upon two vessels, employing the interval to fill one while the water is discharged from the other.

On comparing this with the contrivance previously suggested by De Caus, it will be observed, that even if De Caus knew the physical agent by which the water was driven upward in the apparatus contrived by him, still it was only a means of causing a vessel of boiling water to empty itself; and before a repetition of the process could be obtained, the vessel should be refilled, and again boiled. In the contrivance of Lord Worcester, on the other hand, the agency of the steam was employed in the same manner as it is in the steam engines of the present day, being generated in one vessel, and used for mechanical purposes in another. Nor must this distinction be regarded as trifling and insignificant, because on it depends the whole practicability of using steam as a mechanical agent. Had its action been confined to the vessel in which it was produced, it never could have been employed for any useful purpose.

### SIR SAMUEL MORLAND, 1683.

(28.) It appears, by a MS. in the Harleian Collection in the British Museum, that a mode of applying steam to raise water was proposed to Louis XIV. by Sir Samuel Morland. It contains, however, nothing more than might have been collected from Lord Worcester's description, and is only curious because of the knowledge the writer appears to have had of the expansion which water undergoes in passing into steam. The following is extracted from the MS.:

"The principles of the new force of fire invented by Chevalier Morland in 1682, and presented to his Most Christian Majesty in 1683:—'Water being converted into vapour by the force of fire, these vapours shortly require a greater space (about 2000 times) than the water before occupied, and sooner than be constantly confined, would split a piece of cannon. But being duly regulated according tc the rules of statics, and by science reduced to measure, weight, and balance, then they bear their load peaceably, (like good horses,) and thus become of great use to mankind, particularly for raising water, according to the following table, which shows the number of pounds that may be raised 1800 times per hour to a height of six inches, by cylinders half filled with water, as well as the different diameters and depths of the said cylinders.' "

### DENIS PAPIN, 1695.

(29.) Denis Papin, a native of Blois in France, and professor of mathematics at Marbourg, had been engaged about this period in the contrivance of a machine in which the atmospheric pressure should be made available as a mecha nical agent, by creating a partial vacuum in a cylinder under a piston. His first attempts were directed to the production of this vacuum by mechanical means, having proposed to apply a water-wheel to work an air-pump, and so maintain the degree of rarefaction required. This, however, would eventually have amounted to nothing more than a mode of transmitting the power of the water-wheel to another engine, since the vacuum produced in this way could only give back the power exerted by the water-wheel diminished by the friction of the pumps; still this would attain the end first proposed by Papin, which was merely to transmit the force of the stream of a river, or a fall of water, to a distant point, by partially exhausted pipes or tubes. He next, however, attempted to produce a partial vacuum by the explosion of gunpowder; but this was found to be insufficient, since so much air remained in the cylinder under the piston, that at least half the power due to a vacuum would have been lost. "I have, therefore," proceeds Papin, "attempted to attain this end by another method. Since water being converted into steam by heat acquires the property of elasticity like air, and may afterward be recondensed so perfectly by cold that there will no longer remain the appearance of elasticity in it, I have thought that it would not be difficult to construct machines in which, by means of a moderate heat, and at a small expense, water would produce that perfect vacuum which has been vainly sought by means of gunpowder."

Papin accordingly constructed the model of a machine, consisting of a small pump, in which was placed a solid piston, and in the bottom of the cylinder under the piston was contained a small quantity of water. The piston being in immediate contact with this water, so as to exclude the atmospheric air, on applying fire to the bottom of the cylinder steam was produced, the elastic force of which raised the piston to the top of the cylinder: the fire being then removed, and the cylinder being cooled by the surrounding air, the steam was condensed and reconverted into water, leaving a vacuum in the cylinder into which the piston was pressed by the force of the atmosphere. The fire being applied and subsequently removed, another ascent and descent were accomplished; and in the same manner the alternate motion of the piston might be continued. Papin described no other form of machine by which this property could be rendered available in practice; but he states generally that the same end may be attained by various forms of machines easy to be imagined.*

## THOMAS SAVERY, 1698.

(30.) The discovery of the method of producing a vacuum by the condensation of steam was reproduced before 1698, by Captain Thomas Savery, to whom a patent was granted in that year for a steam engine to be applied to the raising of water, &c. Savery proposed to combine the machine described by the Marquis of Worcester, with an apparatus

* Recueil de diverses pièces touchant quelques nouvelles machines, p. 38.

for raising watei by suction into a vacuum produced by the condensation of steam.

Savery appears to have been ignorant of the publication of Papin, in 1695, and states that his discovery of the condensing principle arises from the following circumstance:—

Having drunk a flask of Florence at a tavern, and flung the empty flask on the fire, he called for a basin of water to wash his hands. A small quantity which remained in the flask began to boil, and steam issued from its mouth. It occurred to him to try what effect would be produced by inverting the flask, and plunging its mouth in the cold water. Putting on a thick glove to defend his hand from the heat, he seized the flask, and the moment he plunged its mouth in the water, the liquid immediately rushed up into the flask and filled it. (21.)

Savery stated that this circumstance immediately suggested to him the possibility of giving effect to the atmospheric pressure by creating a vacuum in this manner. He thought that if, instead of exhausting the barrel of a pump by the usual laborious method of a piston and sucker, it was exhausted by first filling it with steam and then condensing the same steam, the atmospheric pressure would force the water from the well into the pump-barrel, and into any vessel connected with it, provided that vessel were not more than about 34 feet above the elevation of the water in the well. He perceived, also, that, having lifted the water to this height, he might use the elastic force of steam in the manner described by the Marquis of Worcester to raise the same water to a still greater elevation, and that the same steam which accomplished this mechanical effect would serve, by its subsequent condensation, to repeat the vacuum, and draw up more water. It was on this principle that Savery constructed the first engine in which steam was ever brought into practical operation.

# CHAPTER III.

## ENGINES OF SAVERY AND NEWCOMEN.

Savery's Engine.—Boilers and their Appendages.—Working Apparatus.—Mode of Operation.—Defects of the Engine.—Newcomen and Cawley.—Atmospheric Engine.—Accidental Discovery of Condensation by Jet.—Potter's Discovery of the Method of working the Valves

(31.) THE steam engine contrived by Savery, like every other which has since been constructed, consists of two parts essentially distinct. The first is that which is employed to generate the steam, which is called the boiler, and the second, that in which the steam is applied as a moving power.

The former apparatus in Savery's engine consists of two strong boilers, sections of which are represented at D and E in fig. 7; D the greater boiler, and E the less. The tubes T and T' communicate with the working apparatus which we shall presently describe. A thin plate of metal R is applied closely to the top of the greater boiler D turning on a centre C, so that by moving a lever applied to the axis C on the outside of the top, the sliding plate R can be brought from the mouth of the one tube to the mouth of the other alternately. This sliding valve is called the *regulator*, since it is by it that the communications between the boiler and two steam vessels (hereafter described) are alternately opened and closed, the lever which effects this being constantly wrought by the hand of the attendant.

Two *gauge pipes* are represented at G, G', the use of which is to determine the depth of water in the boiler. One G has its lower aperture a little above the proper depth, and the other G' a little below it. Cocks are attached to the upper ends G, G', which can be opened or closed at pleasure. The steam collected in the top of the boiler pressing on the surface of the water forces it up in the tubes G, G', if their lower ends be immersed. Upon opening the cocks G, G', if water be forced from them, there is too much water in the boiler, since the mouth of G is *below* its level. If steam issue from both, there is too little water in the boiler, since the mouth of G' is *above* its level. But if steam issue from G and water from G', the water in the boiler is at its proper level. This ingenious contrivance for determining the level of the water in the boiler is the invention of Savery, and is used in many instances at the present day.

The mouth of G should be at a level of a little less than one-third of the whole depth, and the mouth of G' at a level a little lower than one-third; for it is requisite that about two-thirds of the boiler should be kept filled with water. The tube I forms a communication between the greater boiler D and the lesser or feeding boiler E, descending nearly to the bottom of it. This communication can be opened and closed at pleasure by the cock K. A gauge pipe is inserted similar to G, G', but extending nearly to the bottom. From this boiler a tube F extends, which is continued to a cistern C, (fig. 8) and a cock is placed at M, which, when opened, allows the water from the cistern to flow into the feeding boiler E, and which is closed when that boiler is filled. The manner in which this cistern is supplied will be described hereafter.

Let us now suppose that the principal boiler is filled to the level between the gauge pipes, and that the subsidiary boiler is nearly full of water, the cock K and the gauge cocks G, G' being all closed. The fire being lighted beneath D and the water boiled, steam is produced and is transmitted through one or other of the tubes T T', to the working apparatus. When evaporation has reduced the water in D below the level of G', it will be necessary to replenish the boiler D. This is effected thus. A fire being lighted beneath the feeding boiler E, steam is produced in it above the surface of the water, which having no escape presses on the surface so as to force it up in the pipe I. The cock K being then opened, the boiling water is forced into the principal boiler D, into which it is allowed to flow until water issues from the gauge cock G'. When this takes place, the cock K is closed, and the fire removed from E until the great boiler again wants replenishing. When the feeding boiling E has been exhausted, it is replenished from the cistern C, (fig. 8,) through the pipe F, by opening the cock M.

(32.) We shall now describe the working apparatus in which the steam is used as a moving power.

Let V V' (fig. 8) be two steam vessels communicating by the tubes T T' (marked by the same letters in fig. 7) with the greater boiler D.

Let S be a pipe, called the *suction pipe,* descending into the well or reservoir from which the water is to be raised, and communicating with each of the steam vessels through tubes D D' by valves A A' which open upward. Let F be a pipe continued from the level of the engine of whatever higher level it is intended to elevate the water. The steam vessels V V' communicate with the *force-pipe* F by valves B B, which open upward, through the tubes E E'. Over the steam vessels and on the force-pipe is placed a small cistern C already mentioned, which is kept filled with cold water from the force-pipe, and from the bottom of which proceeds a pipe terminated with a cock G. This is called the *condensing pipe,* and can be brought alternately over each steam vessel. From this cistern another pipe communicates with the feeding boiler (fig. 7) by the cock M.*

The communication of the pipes T T' with the boiler can be opened and closed, alternately, by the regulator R, (fig. 7,) already described.

* This pipe in fig. 9 is represented as proceeding from the force-pipe above the cistern C.

Now suppose the steam vessels and tubes to be all filled with common atmospheric air, and that the regulator be placed so that the communication between the tube T and the boiler be opened, the communication between the other tube T′ and the boiler being closed, steam will flow into V through T. At first, while the vessel V is cold, the steam will be condensed, and will fall in drops of water on the bottom and sides of the vessel. The continued supply of steam from the boiler will at length impart such a degree of heat to the vessel V that it will cease to condense it. Mixed with the heated air contained in the vessel V, it will have an elastic force greater than the atmospheric pressure, and will therefore force open the valve B, through which a mixture of air and steam will be driven until all the air in the vessel V will have passed out, and it will contain nothing but the pure vapour of water.

When this has taken place, suppose the regulator be moved so as to close the communication between the tube T and the boiler, and to stop the further supply of steam to the vessel V; and at the same time let the condensing pipe G be brought over the vessel V, and the cock opened so as to let a stream of cold water flow upon it. This will cool the vessel V, and the steam with which it is filled will be condensed and fall in a few drops of water, leaving the interior of the vessel a vacuum. The valve B will be kept closed by the atmospheric pressure. But the elastic force of the air between the valve A and the surface of the water in the well or reservoir, will open A, so that a part of this air will rush in (6), and occupy the vessel V. The air in the suction pipe S, being thus allowed an increased space, will be proportionably diminished in its elastic force (6), and its pressure will no longer balance that of the atmosphere acting on the external surface L* of the water in the reservoir. This pressure will, therefore, force water up in the tube S until its weight,

* Not in the diagram.

together with the elastic force of the air above it, balances the atmospheric pressure on L, (7.) When this has taken place, the water will cease to ascend.

Let us now suppose that, by shifting the regulator, the communication is opened between T and the boiler, so that steam flows again into V. The condensing cock G being removed, the vessel will be again heated as before, the air expelled, and its place filled by the steam. The condensing pipe being again allowed to play upon the vessel V, and the further supply of steam being stopped, a vacuum will be produced in V, and the atmospheric pressure on L will force the water through the valve A into the vessel V, which it will nearly fill, a small quantity of air, however, remaining above it.

Thus far the mechanical agency employed in elevating the water is the atmospheric pressure; and the power of steam is no further employed than in the production of a vacuum. But, in order to continue the elevation of the water through the force-pipe F, above the level of the steam vessel, it will be necessary to use the elastic pressure of the steam. The vessel V is now nearly filled by the water which has been forced into it by the atmosphere. Let us suppose that, the regulator being shifted again, the communication between the tube T and the boiler is opened, the condensing cock removed, and that steam flows into V. At first coming in contact with the cold surface of the water and that of the vessel, it is condensed; but the vessel is soon heated, and the water formed by the condensed steam collects in a sheet or film on the surface of the water in V, so as to form a surface as hot as boiling water.* The steam then, being no longer condensed, presses on the surface of the water with its elastic force; and when that pressure becomes greater than the atmospheric pressure, the valve B is forced open, and the water, issuing through it, passes through E into the

* Hot water being lighter than cold, it floats on the surface.

force-pipe F; and this is continued until the steam has forced all the water from V, and occupies its place.

The further admission of steam through T is once more stopped by moving the regulator; and the condensing pipe being again allowed to play on V, so as to condense the steam which fills it, produces a vacuum. Into this vacuum, as before, the atmospheric pressure on L will force the water, and fill the vessel V. The condensing pipe being then closed and steam admitted through T, the water in V will be forced by its pressure through the valve B and tube E into F, and so the process is continued.

We have not yet noticed the other steam vessel V′, which, as far as we have described, would have remained filled with common atmospheric air, the pressure of which on the valve A′ would have prevented the water raised in the suction pipe S from passing through it. However, this is not the case; for, during the entire process which has been described in V, similar effects have been produced in V′, which we have only omitted to notice, to avoid the confusion which the two processes might produce. It will be remembered, that after the steam, in the first instance, having flowed from the boiler through T, has blown the air out of V through B, the communication between T and the boiler is closed. Now the same motion of the regulator which closes this opens the communication between T′ and the boiler; for the sliding plate R (fig. 7) is moved from the one tube to the other, and at the same time, as we have already stated, the condensing pipe is brought to play on V. While, therefore, a vacuum is being formed in V by condensation, the steam, flowing through T′, blows out the air through B, as already described in the other vessel V; and, while the air in S is rushing up through A into V, followed by the water raised in S by the atmospheric pressure on L, the vessel V′ is being filled with steam, and the air is completely expelled from it.

The communication between T and the boiler is now

again opened, and the communication between T and the boiler closed by moving the regulator R (fig. 7) from the tube T to T′; at the same time the condensing pipe is removed from over V, and brought to play upon V′. While the steam once more expels the air from V through B, a vacuum is formed by condensation in V′, into which the water in S rushes through the valve A′. In the mean time V is again filled with steam. The communication between T and the boiler is now closed, and that between T′ and the boiler is opened, and the condensing pipe removed from V′, and brought to play on V. While the steam from the boiler forces the water in V′ through B′ into the force-pipe F, a vacuum is being produced in V, into which water is raised by the atmospheric pressure at L.

Thus each of the vessels V V′ is alternately filled from S, and the water thence forced into F. The same steam which forces the water from the vessels into F, having done its duty, is condensed, and brings up the water from S by giving effect to the atmospheric pressure.

During this process, two alternate motions or adjustments must be constantly made; the communication between T and the boiler must be opened, and that between T′ and the boiler closed, which is done by one motion of the regulator. The condensing pipe at the same time must be brought from V to play on V′, which is done by the lever placed upon it. Again the communication between T′ and the boiler is to be opened, and that between T and the boiler closed; this is done by moving back the regulator. The condensing pipe is brought from V′ to V by moving back the other lever, and so on alternately.

For the clearness and convenience of description, some slight and otherwise unimportant changes have been made in the position of the parts.* A perspective view of this

* In the diagrams used for explaining the principles and operations of machines, I have found it contribute much to the clearness of the description t

engine is presented in fig. 9. The different parts already described will easily be recognised, being marked with the same letters as in figs. 6, 7.

(33.) In order duly to appreciate the value of improvements, it is necessary first to perceive the defects which these improvements are designed to remove. Savery's steam engine, considering how little was known of the value and properties of steam, and how low the general standard of mechanical knowledge was in his day, is certainly highly creditable to his genius. Nevertheless it had very considerable defects, and was finally found to be inefficient for the most important purposes to which he proposed applying it.

At the time of this invention, the mines in England had greatly increased in depth, and the process of draining them had become both expensive and difficult; so much so, that it was found in many instances that their produce did not cover the cost of working them. The drainage of these mines was the most important purpose to which Savery proposed to apply his steam engine.

It has been already stated that the pressure of the atmosphere amounts to about 15 lbs. (3) on every square inch. Now, a column of water, whose base is one square inch, and whose height is 34 feet, weighs about 15lbs. If we suppose that a perfect vacuum were produced in the steam vessels v v′ (fig. 8) by condensation, the atmospheric pressure on L would fail to force up the water, if the height of the top ot these vessels exceeded 34 feet. It is plain, therefore, that the engine cannot be more than 34 feet above the water which it is intended to elevate. But in fact it cannot be so much; for the vacuum produced in the steam vessels v v is never perfect. Water, when not submitted to the pressure

adopt an arrangement of parts somewhat different from that of the real machine When once the nature and principles on which the machine acts are well under stood, the reader will find no difficulty in transferring every part to its propei place, which is represented in the perspective drawings.

of the atmosphere, will vaporize at a very low temperature, (17); and it was found that a vapour possessing a considerable elasticity would, notwithstanding the condensation, remain in the vessels V V' and the pipe S, and would oppose the ascent of the water. In consequence of this, it was found that the engine could never be placed with practical advantage at a greater height than 26 feet above the level of the water to be raised.

(34.) When the water is elevated to the engine, and the steam vessels filled, if steam be introduced above the water in V, it must first balance the atmospheric pressure, before it can force the water through the valve B. Here, then, is a mechanical pressure of 15lbs. per square inch expended, without any water being raised by it. If steam of twice that elastic force be used, it will elevate a column in F of 34 feet in height; and if steam of triple the force be used, it will raise a column of 68 feet high, which, added to 26 feet raised by the atmosphere, gives a total lift of 94 feet.

In effecting this, steam of a pressure equal to three times that of the atmosphere acts on the inner surface of the vessels V V'. One third of this bursting of the pressure is balanced by the pressure of the atmosphere on the external surface of the vessels; but an effective pressure of 30lbs. per square inch still remains, tending to burst the vessels. It was found that the apparatus could not be constructed to bear more than this with safety; and, therefore, in practice, the lift of such an engine was limited to about 90 perpendicular feet. In order to raise the water from the bottom of the mine by these engines, therefore, it was necessary to place one at every 90 feet of the depth; so that the water raised by one through the first 90 feet should be received in a reservoir, from which it was to be elevated the next 90 feet by another, and so on.

Besides this, it was found that sufficient strength could not be given to those engines, if constructed upon a large scale. They were, therefore, necessarily very limited in their dimensions, and were incapable of raising the water with sufficient speed. Hence arose a necessity for several engines at each level, which greatly enhanced the expense.

(35.) These, however, were not the only defects of Savery's engines. The consumption of fuel was enormous, the proportion of heat wasted being much more than what was used in either forcing up the water, or producing a *vacuum*. This will be very easily understood by attending to the process of working the engine already described.

When the steam is first introduced from the boiler into the steam vessels V V', preparatory to the formation of a vacuum, it is necessary that it should heat these vessels up to the temperature of the steam itself; for until then the steam will be condensed the moment it enters the vessel by the cool surface. All this heat, therefore, spent in raising the temperature of the steam vessels, is wasted. Again, when the water has ascended and filled the vessels V V', and steam is introduced to force this water through E E' into F, it is immediately condensed by the cold surface in V V', and does not begin to act until a quantity of hot water, formed by condensed steam, is collected on the surface of the cold water which fills the vessel V V'. Hence another source of the waste of heat arises.

When the steam begins to act upon the surface of the water in V V', and to force it down, the cold surface of the vessel is gradually exposed to the steam, and must be heated while the steam continues its action; and when the water has been forced out of the vessel, the vessel itself has been heated to the temperature of the steam which fills it, all which heat is dissipated by the subsequent process of condensation. It must thus be evident that the steam used in forcing up the water in F, and in producing a vacuum, bears a very small proportion indeed to what is consumed in heating the apparatus after condensation.

Pl. 1.

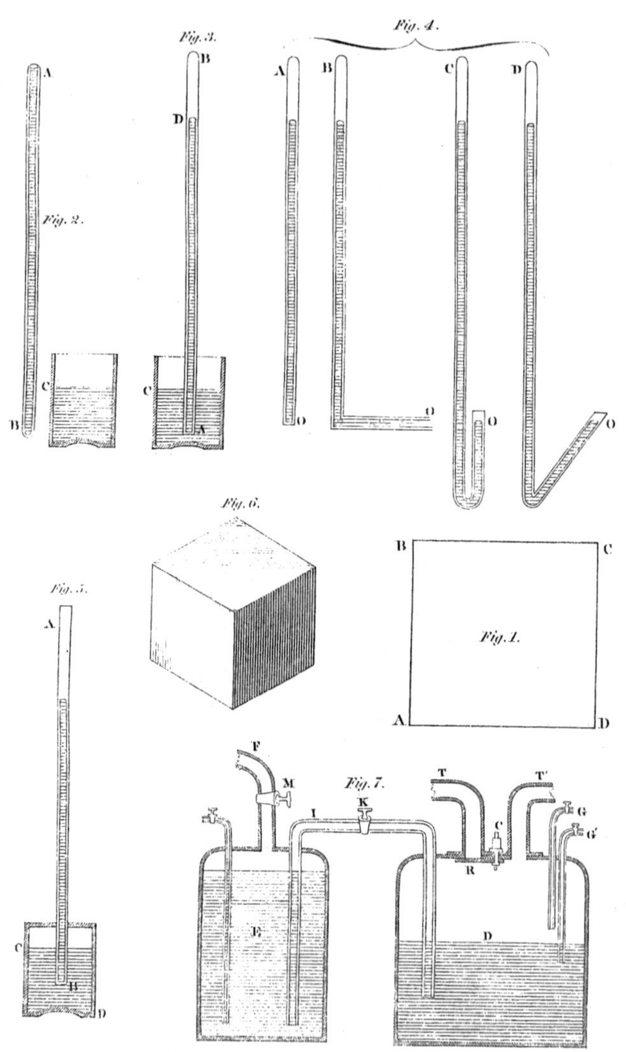


Engd. by P. Maverick.

(36.) There is also another circumstance which increases the consumption of fuel. The water must be forced through n, not only against the atmospheric pressure, but also against a column of 68 feet of water. Steam is therefore required of a pressure of 45lbs. on the square inch. Consequently the water in the boiler must be boiled under this pressure. That this should take place, it is necessary that the water should be raised to a temperature considerably above 212° (17), even so high as 267°; and thus an increased heat must be given to the boiler. Independently of the other defects, this intense heat weakened and gradually destroyed the apparatus.

Besides the drainage of mines, Savery proposed to apply his steam engine to a variety of other purposes; such as supplying cities with water, forming ornamental water-works n pleasure grounds, turning mills, &c.

Savery was the first who suggested the method of expressing the power of an engine with reference to that of horses In this comparison, however, he supposed each horse to work but eight hours a day, while the engine works for 24 hours. This method of expressing the power of steam engines will be explained hereafter.

(37.) The failure of the engines proposed by Captain Savery in the great work of drainage, from the causes which have been just mentioned, and the increasing necessity for effecting this object, arising from the circumstance of the large property in mines, which became every year unproductive by it, stimulated the ingenuity of mechanics to contrive some means of rendering those powers of steam exhibited in Savery's engine practically available. Among others, Thomas Newcomen, a blacksmith of Dartmouth, and John Cawley, a plumber of the same place, turned their attention to this inquiry.

Newcomen appears to have resumed the old method of raising the water from the mines by ordinary pumps, but conceived the idea of working these pumps by some moving power less expensive than that of horses. The means whereby he proposed effecting this, was by connecting the end of a pump rod D (fig. 10) by a chain, with the arch head A of a working beam A B, playing on an axis C. The other arch head B of this beam was connected by a chain with the rod E of a solid piston P, which moved air-tight in a cylinder F. If a vacuum be created beneath the piston P, the atmospheric pressure acting upon it will press it down with a force of 15 lbs. per square inch; and the end A of the beam being thus raised, the pump-rod D will be drawn up. If a pressure equivalent to the atmosphere be then introduced below the piston, so as to neutralize the downward pressure, the piston will be in a state of indifference as to rising or falling; and if in this case the rod D be made heavier than the piston and its rod, so as to overcome the friction, &c., it will descend, and elevate the piston again to the top of the cylinder. The vacuum being again produced, another descent of the piston, and consequent elevation of the pump-rod, will take place; and so the process may be continued.

Such was Newcomen's first conception of the *atmospheric engine;* and the contrivance had much, even at the first view, to recommend it. The power of such a machine would depend entirely on the magnitude of the piston; and being independent of a highly elastic steam, would not expose the materials to the destructive heat which was necessary for working Savery's engine. Supposing a perfect vacuum to be produced under the piston in the cylinder, an effective downward pressure would be obtained, amounting to 15 times as many pounds as there are square inches in the section of the piston.* Thus, if the base of the piston

* As the calculation of the power of an engine depends on the number of square inches in the section of the piston, it may be useful to give a rule for computing the number of square inches in a circle. The following rule will always give the dimensions with sufficient accuracy:—*Multiply the number of inches in the diameter by itself; divide the product by* 14, *and multiply the quotient thus obtained by* 11, *and the result will be the number of square*

were 100 square inches, a pressure equal to 1500 lbs. would be obtained.

(38.) In order to accomplish this design, two things were necessary: 1. To make a speedy and effectual vacuum below the piston in the descent; and 2. To contrive a counterpoise for the atmosphere in the ascent.

The condensation of steam immediately presented itself as the most effectual means of accomplishing the former; and the elastic force of the same steam previous to condensation an obvious method of effecting the latter. Nothing now remained to carry the design into execution, but the contrivance of means for the alternate introduction and condensation of the steam; and Newcomen and Cawley were accordingly granted a patent in 1707, in which Savery was united, in consequence of the principle of condensation, for which he had previously received a patent, being necessary to the projected machine. We shall now describe the *atmospheric engine*, as first constructed by Newcomen:—

The boiler K is placed over a furnace I, the flue of which winds round it, so as to communicate heat to every part of the bottom of it. In the top, which is hemispherical, two gauge pipes G G′ are placed, as in Savery's engine, and a *puppet valve* V, which opens upward, and is loaded at one pound per square inch; so that when the steam produced in the boiler exceeds the pressure of the atmosphere by more than one pound on the square inch, the valve V is lifted, and the steam escapes through it, and continues to escape until its pressure is sufficiently diminished, when the valve V again falls into its seat.

The great steam-tube is represented at S, which conducts

*inches in the circle.* Thus if there be 12 inches in the diameter, this multiplied by itself gives 144, which divided by 14 gives $10\frac{4}{14}$, which multiplied by 11 gives 115, neglecting fractions. There are, therefore, 115 square inches in a circle whose diameter is 12 inches.

steam from the boiler to the cylinder; and a feeding pipe T furnished with a cock, which is opened and closed at pleasure, proceeds from a cistern L to the boiler. By this pipe the boiler may be replenished from the cistern, when the gauge cock G' indicates that the level has fallen below it. The cistern L is supplied with hot water by means which we shall presently explain.

(39.) To understand the mechanism necessary to work the piston, let us consider how the supply and condensation of steam must be regulated. When the piston has been forced to the bottom of the cylinder by the atmospheric pressure acting against a vacuum, in order to balance that pressure, and enable it to be drawn up by the weight of the pump-rod, it is necessary to introduce steam from the boiler. This is accomplished by opening the cock R in the steam-pipe S. The steam being thus introduced from the boiler, its pressure balances the action of the atmosphere upon the piston, which is immediately drawn to the top of the cylinder by the weight of the pump-rod D. It then becomes necessary to condense this steam, in order to produce a vacuum. To accomplish this the further supply of steam must be cut off, which is done by closing the cock R. The supply of steam from the boiler being thus suspended, the diffusion of cold water on the external surface of the cylinder becomes necessary to condense the steam within it. This was done by enclosing the cylinder within another, leaving a space between them.* Into this space cold water is allowed to flow from a cock M placed over it, which is supplied by a pipe from the cistern N. This cistern is supplied with water by a pump O, which is worked by the engine itself, from the beam above it.

The cold water supplied from M, having filled the space between the two cylinders, abstracts the heat from the inner one; and condensing the steam, produces a vacuum into

* The external cylinder is not represented in the diagram.

SAVERY'S STEAM ENGINE

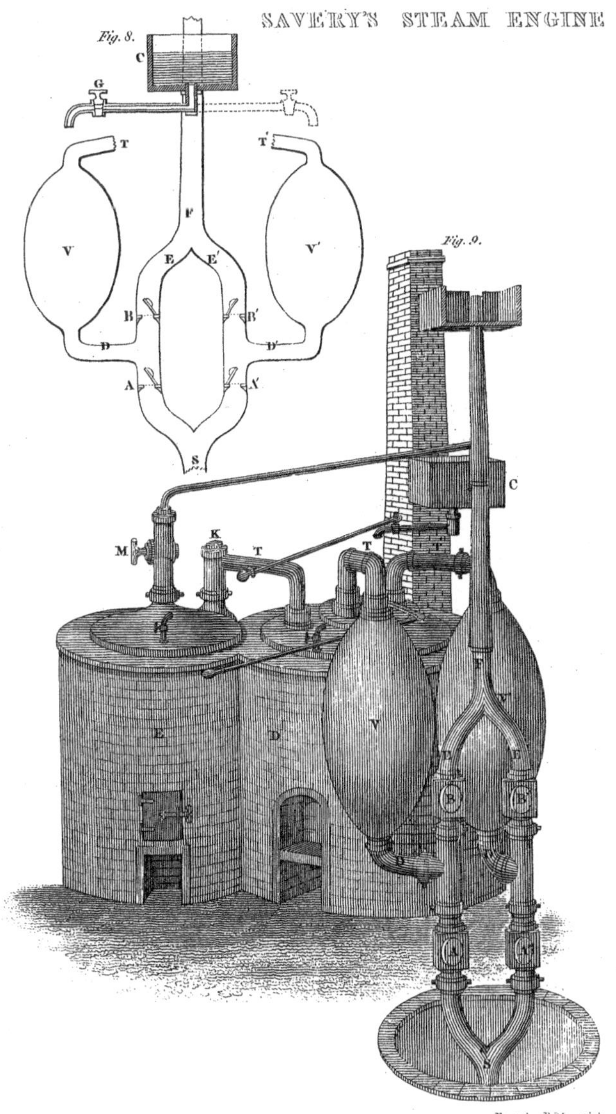


Engr. by P. Maverick.

which the piston is immediately forced by the atmospheric pressure. Preparatory to the next descent, the water which thus fills the space between the cylinders, and which is warmed by the heat it has abstracted from the steam, must be discharged, in order to give room for a fresh supply of cold water from M. An aperture, furnished with a cock, is accordingly provided in the bottom of the cylinder, through which the water is discharged into the cistern L; and being warm, is adapted for the supply of the boiler through T, as already mentioned.

The cock R being now again opened, steam is admitted below the piston, which, as before, ascends, and the descent is again accomplished by opening the cock M, admitting cold water between the cylinders, and thereby condensing the steam below the piston.

The condensed steam, thus reduced to water, will collect in the bottom of the cylinder, and resist the descent of the piston. It is, therefore, necessary to provide an exit for it, which is done by a valve opening *outwards* into a tube which leads to the feeding cistern L, into which the condensed steam is driven.

That the piston should continue to be air-tight, it was necessary to keep a constant supply of water over it; this was done by a cock similar to M, which allowed water to flow from the pipe M on the piston.

(40.) Soon after the first construction of these engines, an accidental circumstance suggested to Newcomen a much better method of condensation than the effusion of cold water on the external surface of the cylinder. An engine was observed to work several strokes with unusual rapidity, and without the regular supply of the condensing water. Upon examining the piston, a hole was found in it, through which the water, which was poured on to keep it air-tight, flowed, and instantly condensed the steam under it.

On this suggestion Newcomen abandoned the external cylinder, and introduced a pipe H furnished with a cock Q into the bottom of the cylinder, so that on turning the cock the pressure of the water in the pipe H, from the level of the water in the cistern N, would force the water to rise as a jet into the cylinder, and would instantly condense the steam. This method of condensing by a jet formed a very important improvement in the engine, and is the method still used.

(41.) Having taken a general view of the parts of the atmospheric engine, let us now consider more particularly its operation.

When the engine is not working, the weight of the pump-rod D draws down the beam A, and draws the piston to the top of the cylinder, where it rests. Let us suppose all the cocks and valves closed, and the boiler filled to the proper depth. The fire being lighted beneath it, the water is boiled until the steam acquires sufficient force to lift the valve V. When this takes place, the engine may be started. For this purpose the regulating valve R is opened. The steam rushes in and is first condensed by the cold cylinder. After a short time the cylinder acquires the temperature of the steam, which then ceases to be condensed, and mixes with the air which filled the cylinder. The steam and heated air, having a greater force than the atmospheric pressure, will open a valve placed at the end X of a small tube in the bottom of the cylinder, and which opens outwards. From this (which is called the *blowing valve**) the steam and air rush in a constant stream until all the air has been expelled, and the cylinder is filled with the pure vapour of water. This process is called *blowing* the engine preparatory to starting it.

When it is about to be started, the engine-man closes the regulator R, and thereby suspends the supply of steam from the boiler. At the same time he opens the *condensing valve* H ;† and thereby throws up a jet of cold water into the cylinder. This immediately condenses the steam contained

* Also called the *snifting valve,* from the peculiar noise made by the air and steam escaping from it.

† Also called the *injection valve.*

in the cylinder, and produces the vacuum. (The atmosphere cannot enter the *blowing valve*, because it opens *outwards*, so that no air can enter to vitiate the vacuum.) The atmospheric pressure above the piston now takes effect, and forces it down in the cylinder. The descent being completed, the engine-man closes the condensing valve H, and opens the regulator R. By this means he stops the play of the jet within the cylinder, and admits the steam from the boiler. The first effect of the steam is to expel the condensing water and condensed steam which are collected in the bottom of the cylinder through the tube Y, containing a valve which opens *outwards*, (called the *eduction valve*,) which leads to the hot cistern L, into which this water is therefore discharged.

When the steam admitted through R ceases to be condensed, it balances the atmospheric pressure above the piston, and thus permits it to be drawn to the top of the cylinder by the weight of the rod D. This ascent of the piston is also assisted by the circumstance of the steam being somewhat stronger than the atmosphere.

When the piston has reached the top, the regulating valve R is closed, and the condensing valve H opened, and another descent produced as before, and so the process is continued.

The manipulation necessary in working this engine was, therefore, the alternate opening and closing of two valves; the regulating and condensing valves. When the piston reached the top of the cylinder, the former was to be closed, and the latter opened; and, on reaching the bottom, the former was to be opened, and the latter closed.

(42.) From the imperfect attention which even an assiduous attendant could give to the management of these valves, the performance of the engines was very irregular, and the waste of fuel very great, until a boy named *Humphrey Potter* contrived means of making the engine work its own valves. This contrivance, although made with no other design than the indulgence of an idle disposition, nevertheless constituted a most important step in the progressive improvement of the steam engine; for by its means, not only the irregularity arising from the negligence of attendants was avoided, but the speed of the engine was doubled.

Potter attached strings to the levers which worked the valves, and carrying these strings to the working beam, fastened them upon it in such a manner that as the beam ascended and descended, it pulled the strings so as to open and close the proper valves with the most perfect regularity and certainty. This contrivance was afterward much improved by an engineer named *Beighton*, who attached to the working beam a straight beam called a *plug frame*, carrying pins which, in the ascent and descent of the beam, struck the levers attached to the valves, and opened and closed them exactly at the proper moment.

The engine thus improved required no other attendance except to feed the boiler occasionally by the cock T, and to attend the furnace.

## CHAPTER IV.

### ENGINE OF JAMES WATT.

Advantages of the Atmospheric Engine over that of Captain Savery.—It contained no new Principle.—Papin's Engine.—James Watt.—Particulars of his Life.—His first Conceptions of the Means of economizing Heat.—Principle of his projected Improvements.

(43.) Considered practically, the engine described in the last chapter possessed considerable advantages over that of Savery; and even at the present day this machine is not unfrequently used in districts where fuel is very abundant and cheap, the first cost being considerably less than that of a modern engine. The low pressure of the steam necessary to work it rendered the use of the atmospheric engine perfectly safe; there being only a bursting pressure of about 1lb. per inch, while in Savery's there was a bursting pressure amounting to 30lbs. The temperature of the steam, not exceeding 216°, did not weaken or destroy the materials; while Savery's engines required steam raised from water at 267°, which in a short time rendered the engine unable to sustain the pressure.

The power of Savery's engines was also very limited, both as to the quantity of water raised, and the height to which it was elevated, (34). On the other hand, the atmospheric engine had no other limit than the dimensions of the piston. In estimating the power of these engines, however, we cannot allow the full atmospheric pressure as an effective force. The condensing water, being mixed with the condensed steam, forms a quantity of hot water in the bottom of the cylinder, which, not being submitted to the atmospheric pressure, (17), produces a vapour which resists the descent of the piston. In practice we find that an allowance of at least 3lbs. per square inch should be made for the resistance of this vapour, and 1lb. per square inch for friction &c.; so that the effective force will be found by subtracting these 4lbs. per square inch from the atmospheric pressure; which, if estimated at 15lbs., leaves an effective working power of about 11lbs. per square inch. This, however, is rather above what is commonly obtained.

Another advantage which this engine has over those of Savery, is the facility with which it might be applied to drive machinery by means of the working beam.

The merit of this engine as an invention must be ascribed principally to its mechanism and combinations. We find in it no new principle; the agency of atmospheric pressure acting against a vacuum, or a partial vacuum, was long known. The formation of a vacuum by the condensation of steam had been suggested by Papin and Savery, and carried into practical effect by the latter. The mechanical power derivable from the direct pressure of the elastic force of steam was distinctly pointed out by Lord Worcester, and even prior to his time; the boiler, gauge pipes, and regulator of the atmospheric engine, were evidently borrowed from Savery's engine. The idea of working a piston in a cylinder by the atmospheric pressure against a vacuum below, was suggested by Otto Guericke, an ingenious German philosopher, the inventor of the air-pump, and subsequently by Papin; and the use of a working beam could not have been unknown. Nevertheless, considerable credit must be acknowledged to be due to Newcomen for the judicious combination of those scattered principles. "The mechanism contrived by him," says Tredgold, "produces all the difference between an efficient and inefficient engine, and should be more highly valued than the fortuitous discovery of a new principle." The rapid condensation of steam by the injection of water, the method of clearing the cylinder of air and water after the stroke, are two contrivances not before in use, and which are quite essential to the effective operation of the engine: these are wholly due to Newcomen and his associates.

(44.) The patent of Newcomen was granted in 1705; and

in 1707, Papin published a work, entitled "A new Method of raising Water by Fire," in which a steam engine is described, which would scarcely merit notice here but for the contests which have arisen upon the claims of different nations for a share in the invention of the steam engine. The publication of this work of Papin was nine years after Savery's patent, with which he acknowledges himself acquainted, and two years after Newcomen's. The following is a description of Papin's steam engine:

An oval boiler A (fig. 11) is filled to about two-thirds of its entire capacity with water, through a valve B in the top, which opens upward, and is kept down by a lever carrying a sliding weight. The pressure on the valve is regulated by moving the weight to or from B, like the common steelyard. This boiler communicates with a cylinder C, by a syphon tube furnished with a stop-cock at D. The cylinder C has a valve F in the top, closed by a lever and weight similar to B, and a tube with a stop-cock G opening into the atmosphere. In this cylinder is placed a hollow copper piston H, which moves freely in it, and floats upon the water. Another tube forms a communication between the bottom of this cylinder and the bottom of a close cylindrical vessel I, called the *air vessel.* In this tube is a valve at K, opening *upward;* also a pipe terminated in a funnel, and furnished with a valve L, which opens *downward.* From the lower part of the air-vessel a tube proceeds, furnished also with a stop-cock M, which is continued to whatever height the water is to be raised.

Water being poured into the funnel, passes through the valve L, which opens downward; and filling the tube, ascends into the cylinder C, carrying the floating piston H on its surface, and maintains the same level in C which it has in the funnel. In this manner the cylinder C may be filled to the level of the top of the funnel. In this process the cock G should be left open, to allow the air in the cylinder to escape as the water rises.

Let us now suppose that, a fire being placed beneath the boiler, steam is being produced. On opening the cock D, and closing G, the steam, flowing through the syphon tube into the top of the cylinder, presses down the floating piston, and forces the water into the lower tube. The passage at L being stopped, since L opens *downward*, the water forces open the valve K, and passes into the air-vessel I. When the piston H has been forced to the bottom of the cylinder, the cock D is closed, and G is opened, and the steam allowed to escape into the atmosphere. The cylinder is then replenished from the funnel as before; and the cock G being closed, and D opened, the process is repeated, and more water forced into the air-vessel I.

By continuing this process, water is forced into the air-vessel, and the air which originally filled that vessel is compressed into the space above the water; and its elastic force increases exactly in the same proportion as its bulk is diminished. (6.) Now, suppose that half of the vessel I has been filled by the water which is forced in, the air above the water being reduced to half its bulk has acquired twice the elastic force, and therefore presses on the surface of the water with twice the pressure of the atmosphere. Again, if two-thirds of the air-vessel be filled with water, the air is compressed into one-third of its bulk, and presses on the surface of the water with three times the pressure of the atmosphere, and so on.

Now, if the cock M be opened, the pressure of the condensed air will force the water up in the tube N, and it will continue to rise until the column balances the pressure of the condensed air. If, when the water is suspended in the tube, and the cock M open, the vessel I is half filled, the height of the column in N will be 34 feet, because 34 feet of water has a pressure equal to the atmosphere; and this, added to the atmospheric pressure on it, gives a total pressure equal to twice that of the atmosphere, which balances the pressure of the air in I reduced to half its bulk. If two-

Pl. III.

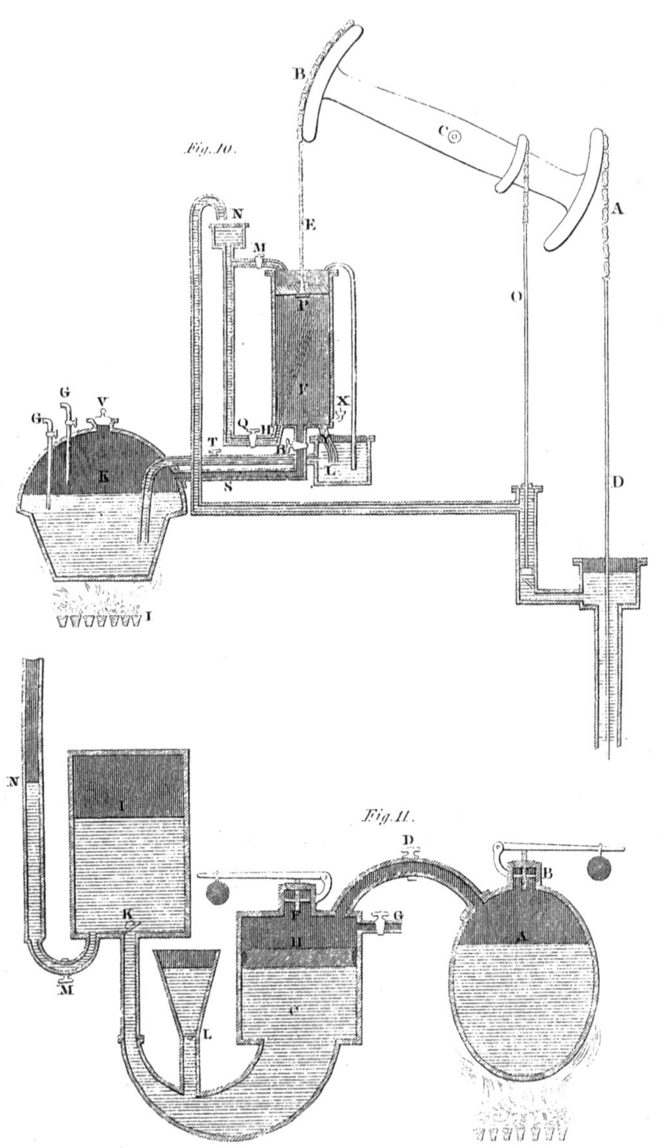


*Drawn by the Author.* *Engd. by P. Maverick.*

nirds of I be filled with water, a column of 68 feet will be supported in N; for such a column, united with the atmospheric pressure on it, gives a total pressure equal to three times that of the atmosphere, which balances the air in I compressed into one-third of its original bulk.

By omitting the principle of condensation, this machine loses 26 feet in the perpendicular lift. But, indeed, in every point of view, it is inferior to the engines of Savery and Newcomen.

(45.) From the construction of the atmospheric engine by Newcomen, in 1705, for about half a century, no very important step had been made in the improvement of the steam engine. During this time the celebrated Smeaton had given much attention to the details of the atmospheric engine, and brought that machine to as high a state of perfection as its principle seemed to admit, and as it has ever since reached.

In the year 1763, JAMES WATT, a name illustrious in the history of mechanical science, commenced his experiments on steam. He was born at Greenock, in the year 1736; and at the age of 16 was apprenticed to a mathematical instrument-maker, with whom he spent four years. At the age of 20 he removed to London, where he still pursued the same trade under a mathematical instrument-maker in that city. After a short time, however, finding his health declining, he returned to Scotland, and commenced business on his own account at Glasgow. In 1757 he was appointed mathematical instrument-maker to the university of Glasgow, where he resided and carried on business.

This circumstance produced an acquaintance between him and the celebrated Dr. Robison, then a student in Glasgow, who directed Watt's attention to the steam engine. In his first experiments he used steam of a high pressure; but found it attended with so much danger of bursting the boiler, and difficulty of keeping the joints tight, and other objections, that he relinquished the inquiry at that time.

(46.) In the winter of 1763, Watt was employed to repair the model of an atmospheric engine, belonging to the natural philosophy class in the university—a circumstance which again turned his attention to the subject of the steam engine. He found the consumption of steam in working this model so great, that he inferred that the quantity wasted must have had a very large proportion to that used in working the piston. His first conclusion was, that the material of the cylinder (brass) was too good a conductor of heat, and that much was thereby lost. He made some experiments, accordingly, with wooden cylinders, soaked in linseed oil, which, however, he soon laid aside. Further consideration convinced him that a prodigious waste of steam was essential to the very principle of the atmospheric engine. This will be easily understood.

When the steam has filled the cylinder so as to balance the atmospheric pressure on the piston, the cylinder must have the same temperature as the steam itself. Now, on introducing the condensing jet, the steam mixed with this water forms a mass of hot water in the bottom of the cylinder. This water, not being under the atmospheric pressure, boils at very low temperatures, and produces a vapour which resists the descent of the piston.

The heat of the cylinder itself assists this process; so that in order to produce a tolerably perfect vacuum, it was found necessary to introduce a quantity of condensing water, sufficient to reduce the temperature of the water in the cylinder lower than 100°, and consequently to cool the cylinder itself to that temperature. Under these circumstances, the descent of the piston was found to suffer very little resistance from any vapour within the cylinder: but then on the subsequent ascent, an immense waste of steam ensued; for the steam, on being admitted under the piston, was immediately condensed by the cold cylinder and water of condensation, and this continued until the cylinder became again heated up to 212°, to which point the whole cylinder should be heated before the ascent could be completed. Here, then, was an obvious and an extensive cause of the waste of heat. At every descent of the piston, the cylinder should be cooled below 100°; and at every ascent it should be again heated to 212°. It, therefore, became a question whether the force gained by the increased perfection of the vacuum was adequate to the waste of fuel in producing the vacuum; and it was found, on the whole, more profitable not to cool the cylinder to so low a temperature, and consequently to work with a very imperfect vacuum, and a diminished power.

Watt, therefore, found the engine involved in this dilemma: either much or little condensation-water must be used. If much were used, the vacuum would be perfect; but then the cylinder would be cooled, and would entail an extensive waste of fuel in heating it. If little were used, a vapour would remain, which would resist the descent of the piston, and rob the atmosphere of a part of its power. The great problem then pressed itself on his attention, *to condense the steam without cooling the cylinder.*

From the small quantity of water in the form of steam which filled the cylinder, and the large quantity of injected water to which this communicated heat, Watt was led to inquire what proportion the bulk of water in the liquid state bore to its bulk in the vaporous state; and also what proportion subsisted between the heat which it contained in these two states. He found by experiment that a cubic inch of water formed about a cubic foot of steam; and that the cubic foot of steam contained as much heat as would raise a cubic inch of water to about 1000°. (15.) This gave him some surprise, as the thermometer indicated the same temperature, 212°, for both the steam and the water from which it was raised. What then became of all the additional heat which was contained in the steam, and not indicated by the thermometer? Watt concluded that this heat must be in some way engaged in maintaining the water in its new form.

Struck with the singularity of this circumstance he com

municated it to Dr. Black, who then explained to Watt his doctrine of *latent heat*, which he had been teaching for a short time before that, but of which Watt had not previously heard; and thus, says Watt, "I stumbled upon one of the material facts on which that theory is founded."

(47.) Watt now gave his whole mind to the discovery of a method of "condensing the steam without cooling the cylinder." The idea occurred to him of providing a vessel separate from the cylinder, in which a constant vacuum might be sustained. If a communication could be opened between the cylinder and this vessel, the steam, by its expansive property, would rush from the cylinder to this vessel, where, being exposed to cold, it would be immediately condensed, the cylinder meanwhile being sustained at the temperature of 212°.

This happy conception formed the first step of that brilliant career which has immortalized the name of Watt, and which has spread his fame to the very skirts of civilization. He states, that the moment the notion of "separate condensation" struck him, all the other details of his improved engine followed in rapid and immediate succession, so that in the course of a day his invention was so complete that he proceeded to submit it to experiment.

His first notion was, as we have stated, to provide a separate vessel, called a *condenser*, having a pipe or tube communicating with the cylinder. This condenser he proposed to keep cold by being immersed in a cistern of cold water, and by providing a jet of cold water to play within it. When the communication with the cylinder is opened, the steam, rushing into the condenser, is immediately condensed by the jet and the cold surface. But here a difficulty presented itself, viz. how to dispose of the condensing water, and condensed steam, which would collect in the bottom of the condenser. But besides this, a quantity of air or permanent uncondensible gas would collect from various sources. Water in its ordinary state always holds more o less air in combination with it: the air thus combined with the water in the boiler passes through the tubes and cylinder with the steam, and would collect in the condenser. Air also would enter in combination with the condensing water, which would be set free by the heat it would receive from admixture with the steam. The air proceeding from these sources would, as Watt foresaw, accumulate in the condenser, even though the water might be withdrawn from it, and would at length resist the descent of the piston. To remedy this, he proposed *to form a communication between the bottom of the condenser and a pump, which he called the* AIR-PUMP, *so that the water and air which might be collected in the condenser would be drawn off;* and it was easy to see how this pump could be worked by the machine itself. This constituted the second great step in the invention.

To make it air-tight in the cylinder, it had been found necessary to keep a quantity of water supplied above the piston. In the present case, any of this water which might escape through the piston, or between it and the cylinder would boil, the cylinder being kept at 212°; and would thus, by the steam it would produce, vitiate the vacuum. To avoid this inconvenience, Watt proposed to lubricate the piston, and keep it air-tight, by employing melted wax and tallow.

Another inconvenience was still to be removed. On the descent of the piston, the air which must then enter the cylinder would lower its temperature; so that upon the next ascent, some of the steam which would enter it would be condensed, and hence would arise a source of waste. To remove this difficulty, Watt proposed to close the top of the cylinder altogether, by an air-tight and steam-tight cover, allowing the piston-rod to play through a hole furnished with a stuffing-box, and *to press down the piston by steam instead of the atmosphere.*

This was the third step in this great invention, and one

which totally changed the character of the machine. It now became really a *steam engine* in every sense; for the pressure above the piston was the elastic force of steam, and the vacuum below it was produced by the condensation of steam; so that steam was used both directly and indirectly as a moving power; whereas, in the atmospheric engine, the indirect force of steam only was used, being adopted merely as an easy method of producing a vacuum.

The last difficulty respecting the economy of heat which remained to be removed, was the circumstance of the cylinder being liable to be cooled on the external surface by the atmosphere. To obviate this, he first proposed casing the cylinder in wood, that being a substance which conducted heat slowly. He subsequently, however, adopted a different method, and enclosed one cylinder within another, leaving a space between them, which he kept constantly supplied with steam. Thus the inner cylinder was kept continually at the temperature of the steam which surrounded it. The outer cylinder was called the *jacket.**

(48.) Watt computed that in the atmospheric engine three times as much heat was wasted in heating the cylinder, &c. as was spent in useful effect. And, as by the improvements proposed by him nearly all this waste was removed, he contemplated, and afterward actually effected, a saving of three-fourths of the fuel.

The honour due to Watt for his discoveries is enhanced by the difficulties under which he laboured from contracted circumstances at the time he made them. He relates, that when he was endeavouring to determine the heat consumed in the production of steam, his means did not permit him to use an efficient and proper apparatus, which would have been attended with expense; and it was by experiments made with apothecaries' phials, that he discovered the property

* It is a remarkable circumstance, that Watt used the same means for keeping the cylinder hot as Newcomen used in his earlier engines to cool it. (38.)

already mentioned, which was one of the facts on which the doctrine of latent heat was founded.

A large share of the merit of Watt's discoveries has, by some writers, been attributed to Dr. Black, to whose instructions on the subject of latent heat it is said that Watt owed the knowledge of those facts which led to his improvements. Such, however, was not the case; and the mistake arose chiefly from some passages respecting Watt in the works of Dr. Robison, in one of which he states that Watt had been a *pupil* and intimate friend of Dr. Black; and that he attended two courses of his lectures at college in Glasgow. Such, however, was not the case; for "Unfortunately for me," says Watt in a letter to Dr. Brewster, "the necessary avocations of my business prevented me from attending his or any other lectures at college. In further noticing Dr. Black's opinion, that his fortunate observation of what happens in the formation and condensation of elastic vapour, 'has contributed in no inconsiderable degree to the public good, by suggesting to my friend Mr. Watt of Birmingham, then of Glasgow, his improvements on the steam engine,' it is very painful for me to controvert any opinion or assertion of my revered friend; yet, in the present case, I find it necessary to say, that he appears to me to have fallen into an error. These improvements proceeded upon the established fact, that steam was condensed by the contact of cold bodies, and the later known one, that water boiled at heats below 100°, and consequently that a vacuum could not be obtained unless the cylinder and its contents were cooled every stroke below the heat."

# CHAPTER V.

## WATT'S SINGLE-ACTING STEAM ENGINE.

Expansive Principle applied.—Failure of Roebuck, and Partnership with Bolton.—Patent extended to 1800.—Counter.—Difficulties in getting the Engines into Use.

(49.) THE first machine in which Watt realized the conceptions which we mentioned in the last chapter, is that which was afterward called his *Single-acting Steam Engine.* We shall now describe the working apparatus in this machine.

The cylinder is represented at C (fig. 12)—in which the piston P moves steam-tight. It is closed at the top, and the piston-rod, being very accurately turned, runs in a steam-tight collar B, furnished with a stuffing-box, and constantly supplied with melted tallow or wax. Through a funnel in the top of the cylinder, melted grease flows upon the piston, so as to maintain it steam-tight. Two boxes A A, containing the valves for admitting and withdrawing the steam, connected by a tube of communication T, are attached to the cylinder; the action of these valves will be presently described. Below the cylinder, placed in a cistern of cold water, is a close cylindrical vessel D, called the condenser, communicating with the cylinder by a tube T', leading to the lower valve-box A. In the side of this condenser is inserted a tube, the inner end of which is pierced with holes, like the rose of a watering-pot; and a cock E in the cold cistern is placed on the outside, through which, when open, the water passing, rises in a jet on the inside.

The tube S, which conducts steam from the boiler, enters the top of the upper valve-box at F. Immediately under it is placed a valve G, which is opened and closed by a lever or rod G'. This valve, when open, admits steam to the top of the piston, and also to the tube T, which communicates between the two valve-boxes, and when closed suspends the admission of steam. There are two valves in the lower box, one H in the top worked by the lever H', and one I in the bottom worked by the lever I'. The valve H, when open, admits steam to pass from the cylinder *above* the piston, by the tube T, to the cylinder *below* the piston, the valve I being supposed in this case to be closed. This valve I, when open, (the valve H being closed,) admits steam to pass from below the cylinder through T' to the condenser. This steam, entering the condenser, meets the jet, admitted to play by the valve E, and is condensed.

The valve G is called the *upper steam valve;* H, *lower steam valve;* I, the *exhausting valve;* and E, the *condensing valve.* Let us now consider how these valves must be worked in order to produce the alternate ascent and descent of the piston.

It is in the first place necessary that all the air which fills the cylinder, tubes, and condenser should be expelled. To accomplish this it is only necessary to open at once the valves G, H, and I. The steam then rushing from F through the valve G will pass into the upper part of the cylinder, and through the tube T and the valve H into the lower part, and also through the valve I into the condenser. After the steam ceases to be condensed by the cold of the apparatus, it will rush out mixed with air through the valve M, which opens outward; and this will continue until all the air has been expelled, and the apparatus filled with pure steam. Then suppose all the valves again closed. The cylinder both above and below the piston is filled with steam; and the steam which filled the condenser being cooled by the cold surface, a vacuum has been produced in that vessel.

The apparatus being in this state, let the upper steam valve G, the exhausting valve I, and the condensing valve E be opened. Steam will thus be admitted through G to press on the top of the piston; and this steam will be prevented from circulating to the lower part of the cylinder by the lower steam-valve H being closed. Also the steam which filled the cylinder below the piston rushes through the open exhausting valve I to the condenser, where it meets the jet allowed to play by the open condensing valve E. It is thus instantly condensed, and a vacuum is left in the cylinder below the piston. Into this vacuum the piston is pressed without resistance by the steam which is admitted through G. When the piston has thus been forced to the bottom of the cylinder, let the three valves G, I, and E, which were before opened, be closed, and let the lower steam-valve H be opened. The effects of this change are easily perceived. By closing the upper steam-valve G, the further admission of steam to the apparatus is stopped. By closing the exhausting valve I, all transmission of steam from the cylinder to the condenser is stopped. Thus the steam which is *in* the cylinder, valve-boxes, and tubes is shut up in them, and no more admitted, nor any allowed to escape. By closing the condensing valve E, the play of the jet in the condenser is suspended.

Previously to opening the valve H, the steam contained in the apparatus was confined to the part of the cylinder *above* the piston and the tube T and the valve-box A. But on opening this valve, the steam is allowed to circulate above and below the piston; and in fact through every part included between the upper steam valve G, and the exhausting valve I. The same steam circulating on both sides, the piston is thus equally pressed upward and downward.

In this case there is no force tending to retain the piston at the bottom of the cylinder except its own weight. Its ascent is produced in the same manner as the ascent of the piston in the atmospheric engine. The piston rod is connected by chains G to the arch-head of the beam, and the weight of the pump-rod R, or any other counterpoise acting on the chains suspended from the other arch-head, draws the piston to the top of the cylinder.

When the piston has arrived at the top of the cylinder, suppose the three valves G, I, and E, to be again opened, and H closed. Steam passes from the steam pipe F through the upper steam valve G to the top of the piston, and at the same time the steam which filled the cylinder below the piston is drawn off through the open exhausting valve I into the condenser, where it is condensed by the jet allowed to play by the open condensing valve E. The pressure of the steam above the piston then forces it without resistance into the vacuum below it, and so the process is continued.

It should be remembered, that of the four valves necessary to work the piston, three are to be opened the moment the piston reaches the top of the cylinder, and the fourth is to be closed; and on the piston arriving at the bottom of the cylinder, these three are to be closed and the fourth opened. The three valves which are thus opened and closed together are the upper steam valve, the exhausting valve, and the condensing valve. The lower steam valve is to be opened at the same instant that these are closed, and *vice versâ*. The manner of working these valves we shall describe hereafter.

The process which has just been described, if continued for any considerable number of reciprocations of the piston, would be attended with two very obvious effects which would obstruct and finally destroy the action of the machine. First, the condensing water and condensed steam would collect in the condenser D, and fill it; and secondly, the water in the cistern in which the condenser is placed would gradually become heated, until at last it would not be cold enough to condense the steam when introduced in the jet. Besides this, it will be recollected that water boils in a vacuum at a very low temperature (17); and, therefore, the hot water collected in the bottom of the condenser would produce steam, which, rising into the cylinder through the exhausting valve, would resist the descent of the piston, and counteract the effects of the steam above it. A further disadvantage arises from the air or other permanently elastic fluid which enters in combination with the water, both in the boiler and condensing jet, and which is disengaged by its own elasticity.

To remove these difficulties, a pump is placed near the condenser, communicating with it by a valve M, which opens from the condenser into the pump. In this pump is placed a piston which moves air-tight, and in which there is a valve N, which opens upward. Now suppose the piston at the bottom of the pump. As it rises, since the valve in it opens *upward*, no air can pass *down* through it, and consequently it leaves a vacuum *below* it. The water and any air which may be collected in the condenser open the valve M, and pass into the lower part of the pump, from which they cannot return in consequence of the valve M opening *outward*. On the descent of the pump piston, the fluids which occupy the lower part of the pump, force open the piston valve N; and passing through it, get *above* the piston, from which their return is prevented by the valve N. In the next ascent, the piston lifts these fluids to the top of the pump, whence they are discharged through a conduit into a small cistern B by a valve K which opens outward. The water which is thus collected in B is heated by the condensed steam, and is reserved in B, which is called the hot well for feeding the boiler, which is effected by means which we shall presently explain. The pump which draws off the hot water and air from the condenser is called the *air-pump*.

(50.) We have not yet explained the manner in which the valves and the air-pump piston are worked. The rod Q of the latter is connected with the working beam, and the pump is therefore wrought by the engine itself. It is not very material to which arm of the beam it is attached. If it be on the same side of the centre of the beam with the cylinder, it rises and falls with the steam piston; but if it be on the opposite side, the pump piston rises when the steam piston falls, and *vice versâ*. In the single engine there are some advantages in the latter arrangement. As the steam piston *descends*, the steam rushes into the condenser, and the jet is playing; and this, therefore, is the most favourable time for drawing out the water and condensed steam from the condenser by the ascent of the pump piston, since by this means the descent of the steam piston is assisted; an effect which would not be produced if the steam piston and pump piston descended together.

With respect to the method of opening and closing the valves, it is evident that the three valves which are simultaneously opened and closed may be so connected as to be worked by the same lever. This lever may be struck by a pin fixed upon the rod Q of the air-pump, so that, when the pistons have arrived at the top of the cylinders, the pin strikes the lever, and opens the three valves. A catch or detent is provided for keeping them open during the descent of the piston, from which they are disengaged in a similar manner on the arrival of the piston at the bottom of the cylinder, and they close by their own weight.

In exactly the same way, the lower steam valve is opened on the arrival of the piston at the bottom of the cylinder, and closed on its arrival at the top, by the action of a pin placed on the piston-rod of the air-pump.

(51.) Soon after the invention of these engines, Watt found that in some instances inconvenience arose from the too rapid motion of the steam piston at the end of its stroke, owing to its being moved with an *accelerated motion.* This was owing to the uniform action of the steam pressure upon it: for upon first putting it in motion at the top of the cylinder, the motion was comparatively slow; but from the continuance of the same pressure, the velocity with which the piston descended was continually increasing, until it reached the bottom of the cylinder, where it acquired its greatest velocity. To prevent this, and to render the descent as nearly as possible uniform, it was proposed to cut off the steam before the descent was completed, so that the remainder might be effected merely by the expansion of the steam which was admitted to the cylinder. To accomplish this, he contrived, by means of a pin on the rod of the air-pump, to close the upper steam valve when the steam piston had completed one-third of its entire descent, and to keep it closed during the remainder of the descent, and until the piston again reached the top of the cylinder. By this arrangement, the steam pressed the piston with its full force through one-third of the descent, and thus put it into motion; during the other two-thirds, the steam thus admitted acted merely by its expansive force, which became less in exactly the same proportion as the space given to it by the descent of the piston increased. Thus, during the last two-thirds of the descent, the piston is urged by a gradually decreasing force, which, in practice, was found just sufficient to sustain in the piston a uniform velocity.

(52.) We have already mentioned the difficulty arising from the water in the cistern, in which the condenser and air-pump are placed, becoming heated, and the condensation therefore being imperfect. To prevent this, a waste pipe is placed in this cistern, from which the water is continually discharged, and a pump L (called the *cold-water pump*) is worked by the engine itself, which raises a supply of cold water, and sends it through a pipe in a constant stream into the cold cistern. The waste pipe, through which the water flows from the cistern, is placed near the top of it, since the heated water, being lighter than the cold, remains on the top. Thus the heated water is continually flowing off, and a constant stream of cold water supplied. The piston-rod of the cold-water pump is attached to the beam (by which it is worked) usually on the opposite side from the cylinder.

Another pump O (called the *hot-water pump*) enters the *hot well* B; and raising the water from it, forces it through a tube to the boiler for the purpose of feeding it. The manner in which this is effected will be more particularly described hereafter. A part of the heat which would otherwise be lost is thus restored to the boiler, to assist in the production of fresh steam. We may consider a portion of the heat to be in this manner *circulating* continually through the machine. It proceeds from the boiler in steam, works the piston, passes into the condenser, and is reconverted into hot water; thence it is passed to the hot well, from whence it is pumped back into the boiler, and is again converted into steam, and so proceeds in constant circulation.

From what has been described, it appears that there are four pistons attached to the great beam, and worked by the piston of the steam cylinder. On the same side of the centre with the cylinder is the piston-rod of the air-pump, and on the opposite side are the piston-rods of the hot-water pump and the cold-water pump; and lastly, at the extremity of the beam opposite to that at which the steam piston works, is the piston of the pump to be wrought by the engine.

(53.) The position of these piston-rods with respect to the centre of the beam depends on the play necessary to be given to the piston. If the play of the piston be short, its rod will be attached to the beam near the centre; and if longer, more remote from the centre. The cylinder of the air-pump is commonly half the length of the steam cylinder, and its piston-rod is attached to the beam at the point exactly in the middle between the end of the beam and the centre. The hot-water pump not being required to raise a considerable quantity of water, its piston requires but little play, and is therefore placed near the centre of the beam, the piston-rod of the cold-water pump being farther from the centre.

(54.) It appears to have been about the year 1763, that Watt made these improvements in the steam engine, and constructed a model which fully realized his expectations.

Either from want of influence, or the fear of prejudice and opposition, he did not make known his discovery, or attempt to secure it by a patent, at that time. Having adopted the profession of a land surveyor, his business brought him into communication with Dr. Roebuck, at that time extensively engaged in mining speculations, who possessed some command of capital, and was of a very enterprising disposition. By Roebuck's assistance and countenance, Watt erected an engine of the new construction at a coal mine on the estate of the Duke of Hamilton, at Kinneil, near Burrowstoness. This engine, being a kind of experimental one, was improved from time to time as circumstances suggested, until it reached considerable perfection. While it was being erected, Watt, in conjunction with Roebuck, applied for and obtained a patent to secure the property in the invention. This patent was enrolled in 1769, six years after Watt invented the improved engine.

Watt was now preparing to manufacture the new engines on an extensive scale, when his partner Roebuck suffered a considerable loss by the failure of a mining speculation in which he had engaged, and became involved in embarrassments, so as to be unable to make the pecuniary advances necessary to carry Watt's designs into execution. Again disappointed, and harassed by the difficulties which he had to encounter, Watt was about to relinquish the further prosecution of his plans, when Mr. Matthew Bolton, a gentleman who had established a factory at Birmingham a short time before, made proposals to purchase Dr. Roebuck's share in the patent, in which he succeeded; and, in 1773, Watt entered into partnership with Bolton.

His situation was now completely changed. Bolton was not only a man of extensive capital, but also of considerable personal influence, and had a disposition which led him, from taste, to undertakings which were great and difficult, and which he prosecuted with the most unremitting ardency and spirit. "Mr. Watt," says Playfair, "was studious and reserved, keeping aloof from the world; while Mr. Bolton was a man of address, delighting in society, active, and mixing with people of all ranks with great freedom, and without ceremony. Had Mr. Watt searched all Europe, he probably would not have found another person so fitted to bring his invention before the public, in a manner worthy of its merit and importance; and although of most opposite habits, it fortunately so happened that no two men ever more cordially agreed in their intercourse with each other."

The delay in the progress of the manufacture of engines occasioned by the failure of Dr. Roebuck was such, that Watt found that the duration of his patent would probably expire before he would even be reimbursed the necessary expenses attending the various arrangements for the manufacture of the engines. He therefore, with the advice and influence of Bolton, Roebuck, and other friends, in 1775, applied to parliament for an extension of the terms of his patent, which was granted for 25 years from the date of his application, so that his exclusive privilege should expire in 1800.

An engine was now erected at Soho (the name of Bolton's factory) as a specimen for the examination of mining speculators, and the engines were beginning to come into demand. The manner in which Watt chose to receive remuneration from those who used his engines was as remarkable for its ingenuity as for its fairness and liberality. He required that one-third of the saving of coals effected by his engines, compared with the atmospheric engines hitherto used, should be paid to him, leaving the benefit of the other two-thirds to the public. Accurate experiments were made to ascertain the saving of coals; and as the amount of this saving in each engine depended on the length of time it was worked, or rather on the number of descents of the piston, Watt invented a very ingenious method of determining this. The vibrations of the great working beam were made to communicate with a train of wheelwork, in the same way as those of a pendulum communicate with the work of a clock. Each vibration of the beam moved one tooth of a small wheel, and the motion was communicated to a hand or index, which moved on a kind of graduated plate like the dial plate of a clock. The position of this hand marked the number of vibrations of the beam. This apparatus, which was called the *counter*, was locked up and secured by two different keys, one of which was kept by the proprietor, and the other by Bolton and Watt, whose agents went round periodically to examine the engines, when the counters were opened by both parties and examined, and the number of vibrations of the beam determined, and the value of the patent third found.*

Notwithstanding the manifest superiority of these engines over the old atmospheric engines; yet such were the influence of prejudice and the dislike of what is new, that Watt found great difficulties in getting them into general use. The comparative first cost also probably operated against them; for it was necessary that all the parts should be executed with great accuracy, which entailed proportionally increased expense. In many instances they felt themselves obliged to induce the proprietors of the old atmospheric engines to replace them by the new ones, by allowing them in exchange an exorbitant price for the old engines; and in some cases they were induced to erect engines at their own expense, upon an agreement that they should only be paid if the engines were found to fulfil the expectations, and brought the advantages which they promised. It appeared since, that Bolton and Watt had actually expended a sum of nearly 50,000*l.* on these engines before they began to receive any

* The extent of the saving in fuel may be judged from this: that for three engines erected at Chacewater mine in Cornwall, it was agreed by the proprietors that they would compound for the patent third at 2400*l.* per annum; so that the whole saving must have exceeded 7200*l.* per annum.

return. When we contemplate the immense advantages which the commercial interests of the country have gained by the improvements in the steam engine, we cannot but look back with disgust at the influence of that fatal prejudice which opposes the progress of improvement under the pretence of resisting innovation. It would be a problem of curious calculation to determine what would have been lost to the resources of this country, if chance had not united the genius of such a man as Watt with the spirit, enterprise, and capital of such a man as Bolton! The result would reflect little credit on those who think novelty alone a sufficient reason for opposition.

---

## CHAPTER VI.

### DOUBLE-ACTING STEAM ENGINE.

The single-acting Engine unfit to impel Machinery.—Various contrivances to adapt it to this purpose.—Double Cylinder.—Double-acting Cylinder.—Various mode of connecting the Piston with the Beam.—Rack and Sector.—Double Chain.—Parallel Motion.—Crank.—Sun and Planet Motion.—Fly Wheel.—Governor.

In the atmospheric engine of Newcomen, and in the improved steam engine of Watt, described in the last chapter, the action of the moving power is an intermitting one. While the piston descends, the moving power is in action, but its action is suspended during the ascent. Thus the opposite or working end of the beam can only be applied in cases where a lifting power is required. This action is quite suitable to the purposes of pumping, which was the chief or only object to which the steam engine had hitherto been applied. In a more extended application of the machine, this intermission of the moving power and its action taking place only in one direction would be inadmissible. To drive the machinery generally employed in manufactures a constant and uniform force is required ; and to render the steam engine available for this purpose, it would be necessary that the beam should be driven by the moving power as well in its ascent as in its descent.

When Watt first conceived the notion of extending the application of the engine to manufactures generally, he proposed to accomplish this double action upon the beam by placing a steam cylinder under each end of it, so that while each piston would be ascending, and not impelled by the steam, the other would be descending, being urged downwards by the steam above it acting against the vacuum below. Thus, the power acting on each during the time when its action on the other would be suspended, a constant force would be exerted upon the beam, and the uniformity of the motion would be produced by making both cylinders communicate with the same boiler, so that both pistons would be driven by steam of the same pressure. One condenser might also be used for both cylinders, so that a similar vacuum would be produced under each.

This arrangement, however, was soon laid aside for one much more simple and obvious. This consisted in the production of exactly the same effect by a single cylinder in which steam was introduced alternately above and below the piston, being at the same time withdrawn by the condenser at the opposite side. Thus the piston being at the top of the cylinder, steam is introduced from the boiler above it, while the steam in the cylinder below it is drawn off by the condenser. The piston, therefore, is pressed from above into the vacuum below, and descends to the bottom of the cylinder. Having arrived there, the top of the cylinder is cut off from all communication with the boiler; and, on the other hand, a communication is opened between it and the condenser. The steam which has pressed the piston down is therefore drawn off by the condenser, while a ccmmunication is opened between the boiler and the bottom o the cylinder, so that steam is admitted below the piston: the piston, thus pressed from below into the vacuum above, ascends, and in the same way the alternate motion is continued. Such is the principle of what is called the *Double-acting Steam Engine,* in contradistinction to that described in the last chapter, in which the steam acts only above the piston, while a vacuum is produced below it.

It is evident that, in the arrangement now described, the condenser must be in constant action: while the piston is descending, the condenser must draw off the steam below it, and while it is ascending, it must draw off the steam above it. As steam, therefore, must be constantly drawn into the condenser, the jet of cold water which condenses the steam must be kept constantly playing. This jet, therefore, will not be worked by the valve alternately opening and closing, as in the single engine, but will be worked by a cock, the opening of which will be adjusted according to the quantity of cold water necessary to condense the steam. When the steam is used at a low pressure, and, therefore, in a less compressed state, less condensing water would be necessary than when it is used at a higher pressure, and in a more compressed state. In the one case, therefore, the condensing cock would be less open than in the other. Again, the quantity of condensing water must vary with the speed of the engine, because, the greater the speed of the engine, the more rapidly will the steam flow from the cylinder into the condenser; and, as the same quantity of steam requires the same quantity of condensing water, the supply of the condensing water must be proportional to the speed of the engine. In the double-acting engine, then, the jet cock is regulated by a lever or index, which moves upon a graduated arch, and which is regulated by the engineer according to the manner in which the engine works.

This change in the action of the steam upon the piston rendered it necessary to make a corresponding change in the mechanism by which the piston-rod was connected with the beam. In the single-acting engine, the piston-rod pulled the end of the beam down during the descent, and was pulled up by it in the ascent. The connexion by which this action was transmitted between the beam and piston was, as we have seen, a flexible chain passing from the end of the piston, and playing upon the arch head of the beam. Now, where the mechanical action to be transmitted is a *pull,* and not a *push,* a flexible chain, or cord, or strap is always sufficient; but if a *push* or *thrust* is required to be transmitted, then the flexibility of the medium of mechanical communication afforded by a chain renders it inapplicable. In the double-acting engine, during the descent, the piston-rod still pulls the beam down, and so far a chain connecting the piston-rod with the beam would be sufficient to transmit the action of the one to the other; but in the ascent the beam no longer pulls up the piston-rod, but is pushed up by it. A chain from the piston-rod to the arch head, as described in the single-acting engine, would fail to transmit this force. If such a chain were used with the double engine, where there is no counter weight on the opposite end of the beam, the consequence would be, that in the ascent of the piston, the chain would slacken, and the beam would still remain depressed. It is therefore necessary that some other mechanical connexion be contrived between the piston-rod and the beam, of such a nature that in the *descent* the piston-rod may *pull* the beam down, and may *push* it up in the *ascent.*

Watt first proposed to effect this by attaching to the end of the piston-rod a straight rack, faced with teeth, which should work in corresponding teeth raised on the arch head of the beam, as represented in fig. 13. If his improved steam engines required no further precision of operation and construction than the atmospheric engines, this might have been sufficient; but in these engines it was indispensably necessary that the piston-rod should be guided with a smooth and even motion through the stuffing box in the top of the cylinder, otherwise any shake or irregularity would cause it to work loose in the stuffing box, and either to admit the air, or to let the steam escape. In fact, it was necessary to turn these piston-rods very accurately in the lathe, so that they may work with sufficient precision in the cylinder. Under these circumstances, the motion of the rack and toothed arch head were inadmissible, since it was impossible by such means to impart to the piston-rod that smooth and equable motion which was requisite. Another contrivance which occurred to Watt was, to attach to the top of the piston-rod a bar which should extend above the beam, and to use two chains or straps, one extending from the top of the bar to the lower end of the arch head, and the other from the bottom of the bar to the upper end of the arch head. By such means, the latter strap would pull the beam down when the piston would descend, and the former would pull the beam up when the piston would ascend. These contrivances, however, were superseded by the celebrated mechanism, since called the *Parallel Motion*, one of the most ingenious mechanical combinations connected with the history of the steam engine.

It will be observed that the object was to connect by some inflexible means the end of the piston-rod with the extremity of the beam, and so to contrive the mechanism, that while the end of the beam would move alternately up and down in a circle, the end of the piston-rod connected with the beam should exactly move up and down in a straight line. If the end of the piston-rod were fastened upon the end of the beam by a pivot, without any other connexion, it is evident that, being moved up and down in the arch of a circle, it would be bent to the left and the right alternately, and would consequently either be broken, or would work loose in the stuffing box. Instead of connecting the end of the rod immediately with the end of the beam by a pivot, Watt proposed to connect them by certain moveable rods, so arranged that, as the end of the beam would move up and down in the circular arch, the rods would so accommodate themselves to that motion, that the end connected with the piston-rod should not be disturbed from its rectilinear course.

To accomplish this, he conceived the notion of connecting three rods, in the following manner:—A B and C D (fig. 14) are two rods or levers, turning on fixed pivots or centres at A and C. A third rod B D, is connected with them by pivots placed at their extremities B and D, and the lengths of the rods are so adjusted that when A B and C D are horizontal, B D shall be perpendicular or vertical, and that A B and C D shall be of equal lengths. Now, let a pencil be imagined to be placed at P, exactly in the middle of the rod B D: if the rod A B be caused to move up and down like the beam of the steam engine in the arch represented in the figure, it is clear, from the mode of their connexion, that the rod C D will be moved up and down in the other arch. Now, Watt conceived that, under such circumstances, the pencil P would be moved up and down in a perpendicular straight line.

However difficult the first conception of this mechanism may have been, it is easy to perceive why the desired effect will be produced by it. When the rod A B rises to the upper extremity of the arch, the point B departs a little to the right; at the same time, the point D is moved a little to the left. Now, the extremities of the rod B D being thus at the same time carried slightly in opposite directions, the pencil in the middle of it will ascend directly upward; the one extremity of the rod having a tendency to draw it as much to the right as the other has to draw it to the left. In the same manner, when the rod A B moves to the lower extremity of the arch, the rod C D will be likewise moved to the lower extremity of its arch. The point B is thus transferred a little to the right, and the point D to the left; and, for the same reason as before, the point P in the middle will move neither to the right nor to the left, but straight downward.*

Now Watt conceived that his object would be attained if he could contrive to make the beam perform the part of A B in fig. 14, and to connect with it other two rods, C D and D B, attaching the end of the piston to the middle of the rod D B. The practical application of this principle required some modification, but is as elegant as the notion itself is ingenious.

The apparatus adopted for carrying it into effect is repre sented on the arm which works the piston in fig. 15. The beam, moving on its axis C, every point in its arm moves in the arc of a circle of which C is the centre. Let B be the point which divides the arm A C into equal parts, A B and B C; and let D E be a straight rod equal in length to C B, and playing on the fixed centre or pivot D. The end E of this rod is connected by a straight bar E B with the point B, by pivots at B and E on which the rod B E plays freely. If the beam be supposed to move alternately on its axis C, the point B will move up and down in a circular arc, of which C is the centre, and at the same time the point E will move in an equal circular arc round the point D as a centre. According to what we have just explained, the middle point F of the rod B E will move up and down in a straight line.

Also, let a rod A G, equal in length to B E, be attached to the end A of the beam by a pivot on which it moves freely, and let its extremity G be connected with E by a rod G E, equal in length to A B, and playing on pivots at G and E.

By this arrangement the joint A G being always parallel

* In a strict mathematical sense, the path of the point F is a curve of a high order, but in the play which is given to it in the application used in the steam engine, it describes only a part of its entire locus; and this part extending equally on each side of a point of inflection, its radius of curvature is infinite, so that, in practice, the deviation from a straight line, when proper proportions are observed in the rods, and too great a play not given to them, is insignificant.

to B E, the three points, C, A, and G will be in circumstances precisely similar to the points C, B, and F, except that the system C A G will be on a scale of double the magnitude of C B F: C A being twice C B, and A G twice B F, it is clear, then, that whatever course the point F may follow, the point G must follow a similar line,* but will move twice as fast. But, since the point F has been already shown to move up and down in a straight line, the point G must also move up and down in a straight line, but of double the length.†

By this arrangement the pistons of both the steam cylinder and air-pump are worked; the rod of the latter being attached to the point F, and that of the former to the point G.

This beautiful contrivance, which is incontestably one of the happiest mechanical inventions of Watt, affords an example with what facility the mind of a mere mechanician can perceive, as it were instinctively, a result to obtain which by strict reasoning would require a very complicated mathematical analysis. Watt, when asked, by persons whose admiration was justly excited by this invention, to what process of reasoning he could trace back his discovery, replied that he was aware of none; that the conception flashed upon his mind without previous investigation, and so as to excite in himself surprise at the perfection of its action; and that on looking at it for the first time, he experienced all that pleasurable sense of novelty which arises from the first contemplation of the results of the invention of others.

* It is, in fact, the principle of the pantograph. The points C, F, and G evidently lie in the same straight line, since C B : C A : : B F : A G, and the latter lines are parallel. Taking C as the common *pole* of the *loci* of the points F G, the *radius vector* of the one will always be twice the corresponding *radius vector* of the other; and therefore these curves are similar, similarly placed, and parallel. Hence, by the last note, the point G must move in a line differing imperceptibly from a right line.

† It is not necessary that the rods, forming the parallel motion, should have the proportions which we have assigned to them. There are various proportions which answer the purpose, and which will be seen by reference to practical works on the steam engine.

This and the other inventions of Watt seem to have been the pure creations of his natural genius, very little assisted by the results of practice, and not at all by the light of education. It does not even appear that he was a dexterous mechanic; for he never assisted in the construction of the first models of his own inventions. His dwelling-house was two miles from the factory, to which he never went more than once in a week, and then did not stay half an hour.

(*a*) However beautiful and ingenious in principle the parallel motion may be, it has recently been shown in the United States that much simpler means are sufficient to subserve the same purpose. In the engines constructed recently, under the direction of Mr. R. L. Stevens, a substitute for the parallel motion has been introduced that performs the task equally well, and is much less complex. On the head of the piston-rod a bar is fixed, at right angles to it, and to the longitudinal section of the engine. The ends of this bar work in guides formed of two parallel and vertical bars of iron, by which the upper end of the piston-rod is constrained to move in a straight line. The cross bar that moves in the guides is connected with the end of the working beam by an inflexible bar, having a motion on two circular gudgeons, one of which is in the working beam, the other in the cross bar. This is therefore free to accommodate itself to the changes in the respective position of the piston-rod and working beam, and yet transmits the power exerted by the steam upon the former, whether it be ascending or descending, to the latter, and through it to the other parts of the machine.—A. E.

(*b*) The most improved form of Watt's engine was reached by successive additions to the old atmospheric engine of Newcomen and Cawley. Hence, the working-beam, derived from the pump brake of that engine, always formed a part; and the parallel motion, or some equivalent contrivance, was absolutely necessary. In many American engines, and particularly in those used in steamboats, the working beam is no longer used for the purpose of transmitting motion to the machinery. This is effected by applying a bar, called the cross head, at right angles to the upper ends of the piston-rod. The ends of the cross head work in iron guides, adapted to a gallows frame of wood. On each side of the cylinder, connecting rods are applied, which take hold of the cranks of the shafts of the water wheel. Two other connecting rods give motion to a short beam, which works the air and supply pumps.

The working beam is also suppressed in engines which work horizontally. The connecting rod is in them merely a jointed prolongation of the piston-rod, extending to the crank, whose axis lies in the same horizontal plane with and at right angles to the axis of the cylinder.—A. E.

(55.) A perfect motion being thus obtained of conveying the alternate motion of the piston to the working beam, the use of a counterpoise to lift the piston was discontinued, and the beam was made to balance itself exactly on its centre. The next end to be obtained was to adapt the reciprocating motion of the working end of the beam to machinery. The motion most generally useful for this purpose is one of *continued rotation.* The object, therefore, was by the *alternate* motion of the end of the beam to transmit to a shaft or axis a *continued circular* motion. In the first instance, Watt proposed effecting this by a *crank,* connected with the working end of the beam by a metal connector or rod.

Let K be the centre or axis, or shaft by which motion is given to the machinery, and to which rotation is to be imparted by the beam C A. On the axle K, suppose a lever K I fixed, so that when K I is turned round the centre K, the wheel must be turned with it. Let a connector or rod, H I, be attached to the points H and I, playing freely on pivots or joints. As the end H is moved upward and downward, the lever K I is turned round the centre K, so as to give a continued rotatory motion to the shaft which revolves on that centre. The different positions which the connector and lever K I assume in the different parts of a revolution are represented in fig. 16.

(56.) This was the first method which occurred to Watt for producing a continued rotatory motion by means of the vibrating motion of the beam, and is the method now universally used. A workman, however, from Mr. Watt's factory, who was aware of the construction of a model of this, communicated the method to Mr. Washborough, of Bristol, who anticipated Watt in taking out a patent; and although it was in his power to have disputed the patent, yet rather than be involved in litigation, he gave up the point, and contrived another way of producing the same effect, which he called the *sun and planet* wheel, and which he used until the expiration of Washborough's patent, when the crank was resumed.

The toothed wheel B (fig. 17) is fixed on the end of the connector, so that it does not turn on its axis. The teeth of this wheel work in those of another wheel A, which is the wheel to which rotation is to be imparted, and which is turned by the wheel B revolving round it, urged by the rod H I, which receives its motion from the working beam. The wheel A is called the sun wheel, and B the planet wheel, from the obvious resemblance to the motion of these bodies.

This contrivance, although in the main inferior to the more simple one of the crank, is not without some advantages; among others, it gives to the sun wheel double the velocity which would be communicated by the simple crank, for in the simple crank one revolution only on the axle is produced by one revolution of the crank, but in the sun and planet wheel two revolutions of the sun wheel are produced by one of the planet wheel; thus a double velocity is obtained from the same motion of the beam. This will be evident from considering that when the planet wheel is in its highest position, its lowest tooth is engaged with the highest tooth of the sun wheel; as the planet wheel passes from the highest position, its teeth drive those of the sun wheel before them, and when it comes into the lowest position, the highest tooth of the planet wheel is engaged with the lowest of the sun wheel: but then half of the sun wheel has *rolled off* the planet wheel, and, therefore, the tooth which was engaged with it in its higher position, must now be distant from it by half the circumference of the wheel, and must, therefore, be again in the highest position, so that, while the planet wheel has been carried from the top to the bottom, the sun wheel has made a complete revolution. A little reflection, however, on the nature of the motion, will render this plainer than any description can. This advantage of giving an increased velocity, may be obtained also by the simple crank, by placing toothed wheels on its axle. Independently of the greater expense attending the construction of the sun and planet wheel, its liability to go out of order, and the rapid wear of the teeth, and other objections, rendered it decidedly inferior to the crank, which has now entirely superseded it.

(58.) Whether the simple crank or the sun and planet wheel be used, there still remains a difficulty of a peculiar nature attending the continuance of the rotatory motion. There are two positions in which the engine can give no motion whatever to the crank. These are when the end of the beam, the axle of the crank, and the pivot which joins the connector with the crank, are in the same straight line. This will be easily understood. Suppose the beam, connector, and crank to assume the position represented in fig. 15. If steam urge the piston downward, the point H and the connector H I will be drawn directly upward. But it must be very evident that in the present situation of the connector H I, and the lever I K, the force which draws the point I in the direction I K can have no effect whatever in turning I K round the centre K, but will merely exert a pressure on the axle or pivots of the wheel.

Again, suppose the crank and connector to be in the position H I K, (fig. 16,) the piston being consequently at the bottom of the cylinder. If steam now press the piston *upward*, the pivot H and the connector H I will be pressed *downward*, and this pressure will urge the crank I K in the direction I K. It is evident that such a force cannot turn the crank round the centre K, and can be attended with no other effect than a pressure on the axle or pivots of the wheel.

Hence, in these two positions, the engine can have no effect whatever in turning the crank. What, then, it may be asked, extricates the machine from this mechanical dilemma, in which it is placed twice in every revolution, on arriving at those positions in which the crank escapes the influence of the power? There is a tendency in bodies, when once put in motion, to continue that motion until stopped by some opposing force, and this tendency carries the crank out of those two critical situations. The velocity which is given to it, while it is under the influence of the impelling force of the beam, is retained in a sufficient degree to carry it through that situation in which it is deserted by this impelling force. Although the rotatory motion intended to be produced by the crank is, therefore, not absolutely destroyed by this circumstance, yet it is rendered extremely irregular, since, in passing through the two positions already described, where the machine loses its power over the crank, the motion will be very slow, and, in the positions of the crank most remote from these, where the power of the beam upon it is greatest, the motion will be very quick. As the crank revolves from each of those positions where the power of the machine over it is greatest, to where that power is altogether lost, it is continually diminished, so that, in fact, the crank is driven by a varying power, and therefore produces a varying motion. This will be easily understood by considering the successive positions of the crank and connector represented in fig. 16.

This variable motion becomes particularly objectionable when the engine is employed to drive machinery. To remove this defect, we have recourse to the property of bodies just mentioned, viz. their tendency to retain a motion which is communicated to them. A large metal wheel called a *fly wheel* is placed upon the axis of the crank, (fig. 15,) and is turned by it. The effect of this wheel is to equalize the motion communicated by the action of the beam on the crank, that action being just sufficient to sustain in the fly wheel a uniform velocity, and the tendency of this wheel to retain the velocity it receives, renders its rotation sufficiently uniform for all practical purposes.

This uniformity of motion, however, will only be preserved on two conditions; *first,* that the supply of steam from the boiler shall be uniform; and, secondly, that the machine have always the same resistance to overcome or be *loaded* equally. If the supply of steam from the boiler to the cylinder be increased, the motion of the piston will be rendered more rapid, and, therefore, the revolution of the fly wheel will also be more rapid, and, on the other hand, a diminished supply of steam will retard the fly wheel. Again, if the resistance or load upon the engine be diminished, the supply of steam remaining the same, the velocity will be increased, since a less resistance is opposed to the energy of the moving power; and, on the other hand, if the resistance or load be increased, the speed will be diminished, since a greater resistance will be opposed to the same moving power. To ensure a uniform velocity, in whatever manner the load or resistance may be changed, it is necessary to proportion the supply of steam to the resistance, so that, upon the least variation in the velocity, the supply of steam will be increased or diminished, so as to keep the engine going at the same rate.

(59.) One of the most striking and elegant appendages of the steam engine is the apparatus contrived by Watt for effecting this purpose. An apparatus, called a *regulator,* or *governor*, had been long known to mill-wrights for rendering uniform the action of the stones in corn mills, and was used generally in machinery. Mr. Watt contrived a beautiful application of this apparatus for the regulation of the steam engine. In the pipe which conducts steam from the boiler to the cylinder he placed a thin circular plate, so that when placed with its face presented toward the length of the pipe, it nearly stopped it, and allowed little or no steam to pass to the cylinder, but when its edge was placed in the direction of the pipe, it offered no resistance whatever to the passage of the steam. This circular plate, called the throttle valve, was made to turn on a diameter as an axis, passing consequently through the centre of the tube, and was worked by a lever outside the tube. According to the position given to it, it would permit more or less steam to pass. If the valve be placed with its edge nearly in the direction of the tube, the supply of steam is abundant; if it be placed with its face nearly in the direction of the tube, the supply of steam is more limited, and it appears that, by the position given to this valve, the steam may be measured in any quantity to the cylinder.

At first it was proposed that the engine man should adjust this valve with his hand; when the engine was observed to increase its speed too much, he would check the supply of steam by partially closing the valve; but if, on the other hand, the motion was too slow, he would open the valve, and let in a more abundant supply of steam. Watt, however, was not content with this, and desired to make the engine itself discharge this task with more steadiness and regularity than any attendant could, and for this purpose he applied the governor already alluded to.

This apparatus is represented in fig. 15; L is a perpendicular shaft or axle to which a wheel M with a groove is attached. A strap or rope, which is rolled upon the axle of the fly wheel, is passed round the groove in the wheel M, in the same manner as the strap acts in a turning lathe By means of this strap the rotation of the fly wheel will produce a rotation of the wheel M and the shaft L, and the speed of the one will always increase or diminish in the same proportion as the speed of the other. N, N are two heavy balls of metal placed at the ends of rods, which play on an axis fixed on the revolving shaft at O, and extend beyond the axis to Q Q. Connected with these by joints at Q Q are two other rods, Q R, which are attached to a broad ring of metal, moving freely up and down the revolving shaft. This ring is attached to a lever whose centre is S, and is connected by a series of levers with the throttle valve T. When the speed of the fly wheel is much increased, the spindle L is whirled round with considerable rapidity, and by their natural tendency* the balls N N fly from the centre. The levers which play on the axis O, by this motive, diverge from each other, and thereby depress the joints Q Q, and draw down the joints R, and with them the ring of metal which slides upon the spindle. By these means, the end of the lever playing on S is depressed, and the end V raised, and the motion is transmitted to the throttle valve, which is thereby partially closed, and the supply of steam to the cylinder checked. If, on the contrary, the velocity of the fly wheel be diminished, the balls will fall toward the axis, and the opposite effects ensuing, the supply of steam will be increased, and the velocity restored.

The peculiar beauty of this apparatus is, that in whatever position the balls settle themselves, the velocity with which the governor revolves must be the same,† and in this, in fact,

* The *centrifugal force.*

† Strictly speaking, this is only true when the divergence of the rods from the spindle is not very great, and, in practice, this divergence is never sufficient to render the above assertion untrue. This property of the conical pendulum arises from the circumstance of the centrifugal force, in this instance, varying as the radius of the circle in which the balls are moved; and when this is the case, as is well known, the *periodic time* is constant. The time of one revolution of the balls is equal to twice the time in which either ball, as a common pendulum, would vibrate on the centre, and as all its vibrations, though the

Pl. IV.

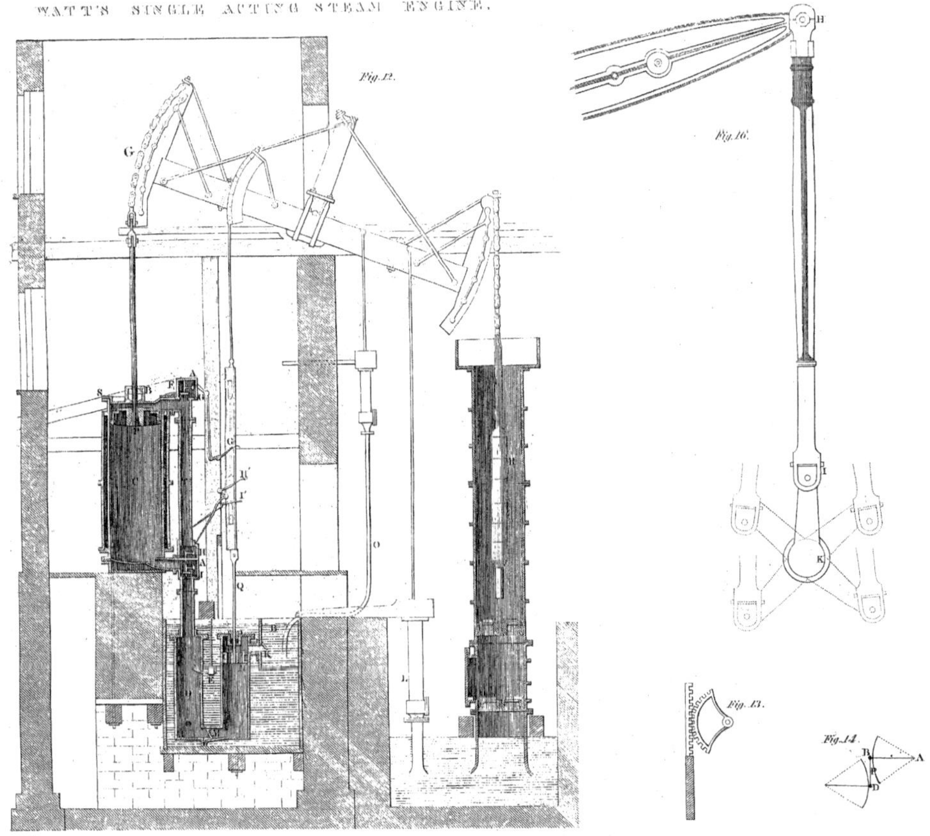


Drawn by the Author

Engr. by P. Maverick.

consists its whole efficacy as a regulator. Its regulating power is limited, and it is only small changes of velocity that it will correct. It is evident that such a velocity as, on the one hand, would cause the balls to fly to the extremity of their play, or, on the other, would cause them to fall down on their rests, would not be influenced by the governor.

We have thus described the principal parts of the double-acting steam engine. The valves and the methods of working them have been reserved for the next chapter, as they admit of considerable variety, and will be better treated of separately. We have also reserved the consideration of the boiler, which is far from being the least interesting part of the modern steam engine, for a future chapter.

arcs be unequal, are equal in time, provided those arcs be small, so also is the periodic time of the revolving ball invariable. These observations, however, only apply when the balls settle themselves steadily into a circular motion; for while they are ascending they describe a spiral curve with double curvature, and the period will vary. This takes place during the momentary changes in the velocity of the engine.

## CHAPTER VII.

### DOUBLE-ACTING STEAM ENGINE.

(CONTINUED.)

On the Valves of the Double-acting Steam Engine.—Original Valves.—Spindle Valves.—Sliding Valve.—D Valve.—Four-way Cock.

(60.) THE various improvements described in the last chapter were secured to Watt by patent in the year 1782. The engine now acquired an enlarged sphere of action; for its dominion over manufactures was decided by the *fly wheel*, *crank*, and *governor*. By means of these appendages, its motions were regulated with the most delicate precision; so that while it retained a power whose magnitude was almost unlimited, that power was under as exact regulation as the motion of a time-piece. There is no species of manufacture, therefore, to which this machine is not applicable, from the power which spins the finest thread, or produces the most delicate web, to that which is necessary to elevate the most enormous weights, or overcome the most unlimited resistances. Although it be true, that in later times the steam engine has received many improvements, some of which are very creditable to the invention and talents of their projectors, yet it is undeniable that all its great and leading perfections, all those qualities by which it has produced such wonderful effects on the resources of these countries, by the extension of manufactures and commerce,—those qualities by which its influence is felt and acknowledged in every part of the civilized globe, in increasing the happiness, in multiplying the enjoyments, and cheapening the pleasures of life,—that these qualities are due to the predominating powers of one man, and that man one who possessed neither the influence of wealth, rank, nor education, to give that first impetus which is so often necessary to carry into circulation the earlier productions of genius.

The method of working the valves of the double-acting steam engine is a subject which has much exercised the ingenuity of engineers, and many elegant contrivances have been suggested, some of which we shall now proceed to describe. But even in this the invention of Watt has anticipated his successors; and the contrivances suggested by him are those which are now almost universally used.

In order perfectly to comprehend the action of the several systems of valves which we are about to describe, it will be necessary distinctly to remember the manner in which the steam is to be communicated to the cylinder, and withdrawn from it. When the piston is at the top of the cylinder, the steam below it is to be drawn off to the condenser, and the steam from the boiler is to be admitted above it. Again, when it has arrived at the bottom of the cylinder, the steam above is to be drawn off to the condenser, and the steam from the boiler is to be admitted below it.

In the earlier engines constructed by Watt, this was accomplished by four valves, which were opened and closed in pairs. Valve boxes were placed at the top and bottom of the cylinder, each of which communicated by tubes both with the steam pipe from the boiler and the condenser. Each valve box accordingly contained two valves, one to admit steam from the steam pipe to the cylinder, and the other to allow that steam to pass into the condenser. Thus each valve box contained a steam valve and an exhausting valve. The valves at the top of the cylinder are called the *upper steam valve* and the *upper exhausting valve*, and those at the bottom, the *lower steam valve* and the *lower exhausting valve*. In fig. 15, A′ is the upper steam valve, which, when open, admits steam above the piston; B′ is the upper exhausting valve, which, when open, draws off the steam from the piston to the condenser. C' is the lower steam valve, which admits steam below the piston; and D, the lower exhausting valve, which draws off the steam from below the piston to the condenser.

Now, suppose the piston to be at the top of the cylinder, the cylinder below it being filled with steam, which has just pressed it up. Let the *upper steam valve* A', and the *lower exhausting valve* D', be opened, and the other two valves closed. The steam which fills the cylinder below the piston will immediately pass through the valve D' into the condenser, and a vacuum will be produced below the piston. At the same time, steam is admitted from the steam pipe through the valve A' above the piston, and its pressure will force the piston to the bottom of the cylinder. On the arrival of the piston at the bottom of the cylinder, the upper steam valve A', and lower exhausting valve D', are closed; and the lower steam valve C', and upper exhausting valve B', are opened. The steam which fills the cylinder above the piston now passes off through B' into the condenser, and leaves a vacuum above the piston. At the same time, steam from the boiler is admitted through the lower steam valve C', below the piston, so that it will press the piston to the top of the cylinder; and so the process is continued.

It appears, therefore, that the upper steam valve, and the lower exhausting valve, must be opened together, on the arrival of the piston at the top of the cylinder. To effect this, one lever, E', is made to communicate by jointed rods with both these valves, and this lever is moved by a pin placed on the piston-rod of the air-pump; and such a position may be given to this pin as to produce the desired effect exactly at the proper moment of time. In like manner, another lever, F', communicates by jointed rods with the upper exhausting valve and lower steam valve, so as to open them and close them together; and this lever, in like manner, is worked by a pin on the piston-rod of the air-pump.

61.) This method of connecting the valves, and working them, has been superseded by another, for which Mr. MURRAY, of Leeds, obtained a patent, which was, however, set aside by Messrs. Bolton and Watt, who showed that they had previously practised it. This method is represented in figs. 18, 19. The stems of the valves are perpendicular, and move in steam-tight sockets in the top of the valve boxes. The stem of the upper steam valve A is a tube through which the stem of the upper exhausting valve B passes, and in which it moves steam-tight; both these stems moving steam-tight through the top of the valve box. The lower steam valve C, and exhausting valve D, are similarly circumstanced; the stem of the former being a tube through which the stem of the latter passes. The stems of the upper steam valve and lower exhausting valve are then connected by a rod E; and those of the upper exhausting valve and lower steam valve by another rod F. These rods, therefore, are capable of moving the valves in pairs, when elevated and depressed. The motion which works the valves is, however, not communicated by the rod of the air-pump, but is received from the axis of the fly wheel. This axis works an apparatus called an *eccentric;* the principle which regulates the motion of this may be thus explained:—

D E (figs. 20, 21) is a circular metallic ring, the inner surface of which is perfectly smooth. This ring is connected with a shaft F D, which communicates motion to the valves by levers which are attached to it at B. A circular metallic plate is fitted in the ring, so as to be capable of turning within it, the surfaces of the ring and plate which are in contact being smooth, and lubricated with oil or grease. This circular plate revolves, but not on its centre. It turns on an axis C, at some distance from its centre A; the effect of which, evidently, is that the ring within which it is turned is moved alternately in opposite directions, and through a space equal to twice the distance (C A) of the axis of the circular plate from the common centre of it and the ring. The eccentric in its two extreme positions is represented in figs. 20, 21. The plate and ring D E are placed on the axis of the fly wheel, or on the axis of some other wheel which is worked by the fly wheel. So that the motion of continued rotation in the fly wheel is thus made to produce an alternate motion in a straight line in the shaft F B. This rod is made to communicate by levers with the rods E and F, (figs. 18, 19,) which work the valves in such a manner, that, when the eccentric is in the position fig. 20, one pair of valves are opened, and the other pair closed; and when it is brought to the position fig. 21, the other pair are opened and the former closed, and so on. It is by means of such an apparatus as this that the valves are worked almost universally at present.

The piston being supposed to be at the top of the cylinder, (fig. 18,) and the rod E raised, the valves A and D are opened, and B and C closed. The steam enters from the steam pipe at an aperture immediately above the valve A, and, passing through the open valve, enters the cylinder above the piston. At the same time, the steam which is below the piston, and which has just pressed it up, flows through the open valve D, and through a tube immediately under it, to the condenser. A vacuum being thus produced below the piston, and steam pressure acting above it, it descends; and when it arrives at the bottom of the cylinder, (fig. 19,) the rod F is drawn down, and the valves A and D fall into their seats, and at the same time the rod F is raised, and the valves B and C are opened. Steam is now admitted through an aperture above the valve C, and passes below the piston, while the steam above it passes through the open valve B into a tube immediately under it, which leads to the condenser. A vacuum being thus produced above the piston, and steam pressure acting below it, the piston ascends, and thus the alternate ascent and descent is continued by the motion communicated to the rods E F from the fly wheel.

WATT'S DOUBLE ACTING STEAM ENGINE

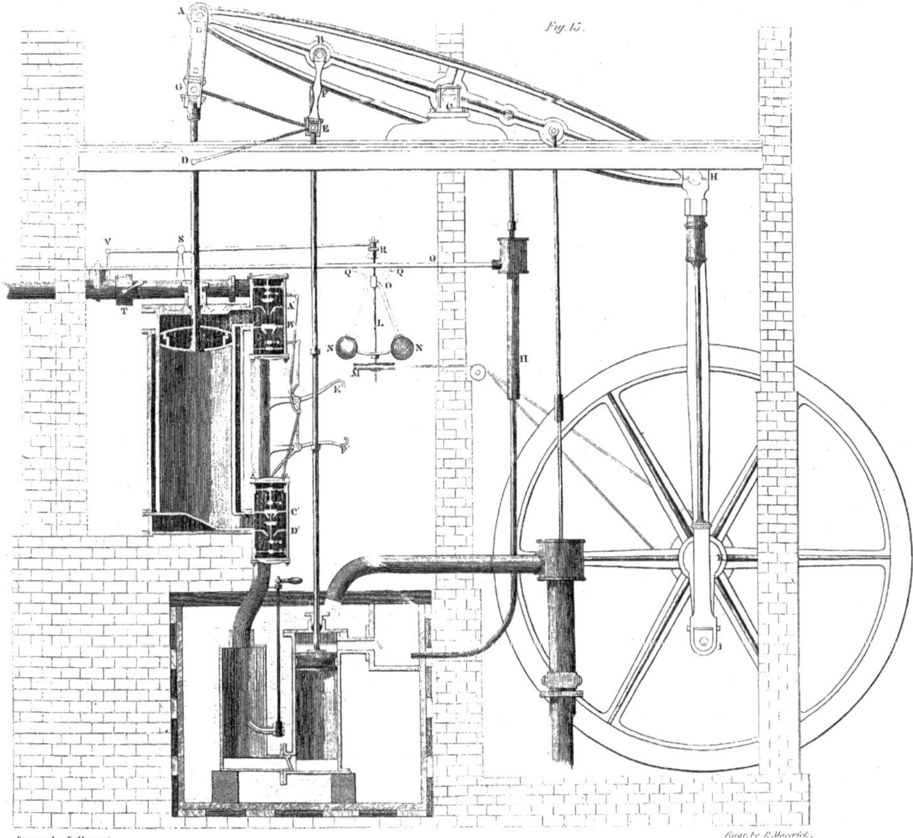


Drawn by J. Clement.

Engr. by E. Maverick.

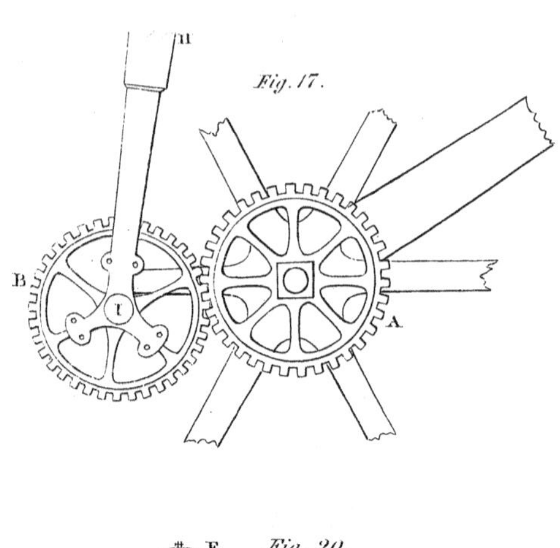


*Fig. 17.*

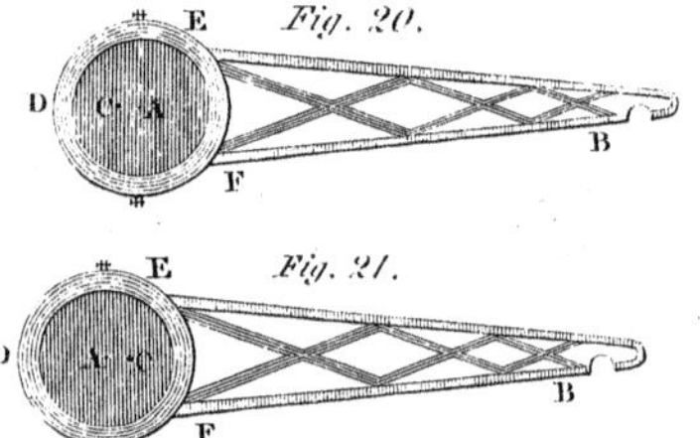


*Fig. 20.*

*Fig. 21.*

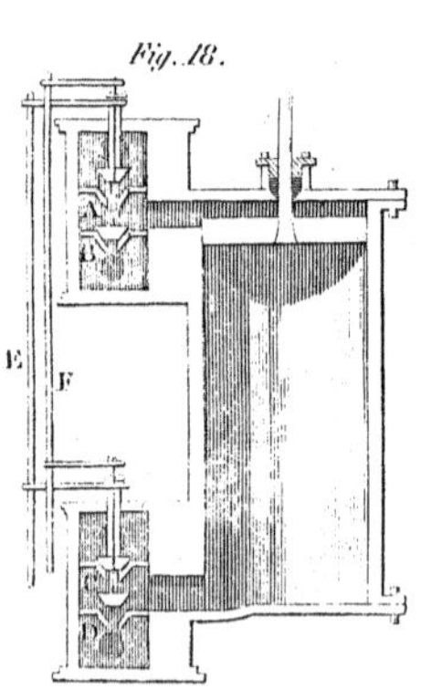


*Fig. 18.*

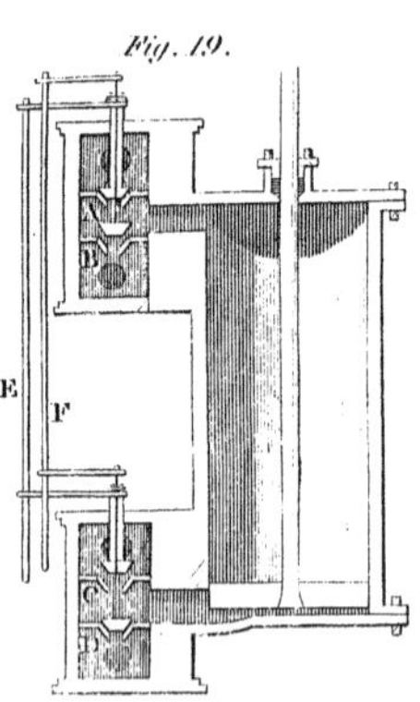


*Fig. 19.*

Drawn by the Author. Engr. by Peter Maverick

(*c*) An improvement has been made in the United States in the mode of working the puppet valve. It consists in placing them by pairs in two different vertical planes instead of one. The rods then work through four separate stuffing boxes, and the necessity of making two of them hollow cylinders is avoided.—A. E.

(62.) There are various other contrivances for regulating the circulation of steam through the cylinder. In figs. 22, 23, is represented a section of a slide valve suggested by Mr. Murray of Leeds. The steam pipe from the boiler enters the valve box D E at S. Curved passages, A A, B B, communicate between this valve box and the top and bottom of the cylinder; and a fourth passage leads to the tube C, which passes to the condenser. A sliding piece within the valve box opens a communication alternately between each end of the cylinder and the tube C, which leads to the condenser. In the position of the apparatus in fig. 22, steam is passing from the steam pipes, through the curved passage A A above the piston, and at the same time the steam below the piston is passing through the passage B B into the tube C, and thence to the condenser. A vacuum is thus forced below the piston, and steam is introduced above it. The piston, therefore, descends; and when it arrives at the bottom of the cylinder, the slide is moved into the position represented in fig. 23. Steam now passes from S through B B below the piston, and the steam above it passes through A A and C to the condenser. A vacuum is thus produced above the piston, and steam pressure is introduced below it, and the piston ascends; and in this way the motion is continued.

The slide is moved by a lever, which is worked by the eccentric from the fly wheel.

(63.) Watt suggested a method of regulating the circulation of steam, which is called the D valve, from the resemblance which the horizontal section of the valve has to the letter D. This method, which is very generally used, is represented in section in figs. 24, 25. Steam from the boiler enters through s. A rod of metal connects two solid plugs, A B, which move steam-tight in the passage D. In the position of the apparatus represented in fig. 24, the steam passes from s through the passage D, and enters the cylinder above the piston; while the steam below the piston passes through the open passage by the tube C to the condenser. A vacuum is thus formed below the piston, while the pressure of steam is introduced above it, and it accordingly descends. When it has arrived at the bottom of the cylinder, the plugs A B are moved into the position in fig. 25. Steam now passing from s through D, enters the cylinder below the piston; while the steam which is above the piston, and has just pressed it down, passes through the open passage into the condenser. A vacuum is thus produced above the piston, and the steam pressure below forces it up. When it has arrived at the top of the cylinder, the position of the plugs A B is again changed to that represented in fig. 24, and a similar effect to that already described is produced, and the piston is pressed down; and so the process is continued.

The plugs A B, and the rod which connects them, are moved up and down by proper levers, which receive their motion from the eccentric.

This contrivance is frequently modified, by conducting the steam from above the piston to the condenser, through a tube in the plugs A B, and their connecting rod. In figs. 26, 27, a tube passes through the plugs A B and the rod which joins them. In the position fig. 26, steam entering at s passes through the tube to the cylinder above the piston, while the steam below the piston passes through C into the condenser. A vacuum being thus made below the piston, and steam pressing above it, it descends; and when it has arrived at the bottom of the cylinder, the position of the plugs A B and the tube is changed to that represented at fig. 27. The steam now entering at s passes to the cylinder below the piston, while the steam above the piston passes through C into the condenser. A vacuum is thus produced above the piston, and steam pressure introduced below it, so that it ascends. When it has arrived at the top of the cylinder, the plugs are moved into the position represented in fig. 26, and similar effects being produced, the piston again descends: and so the motion is continued.

The motion of the sliding tube may be produced as in the former contrivances, by the action of the eccentric. It is also sometimes done by a bracket fastened on the piston-rod of the air-pump. This bracket, in the descent of the piston, strikes a projection on the valve-rod, and drives it down; and in the ascent meets a similar projection, and raises it.

(64.) Another method, worthy of notice for its elegance and simplicity, is the *four-way cock.* A section of this contrivance is given in figs. 28, 29; C T S B are four passages or tubes; S leads from the boiler, and introduces steam; C, opposite to it, leads to the condenser; T is a tube which communicates with the top of the cylinder; and B one which communicates with the bottom of the cylinder. These four tubes communicate with a cock, which is furnished with two curved passages, as represented in the figures; and these passages are so formed, that, according to the position given to the cock, they may be made to open a communication between any two adjacent tubes of the four just mentioned. When the cock is placed as in fig. 28, communication is opened between the steam pipe and the top of the cylinder by one of the curved passages, and between the condenser and the bottom of the cylinder by the other curved passage. In this case the steam passes from below the piston to the condenser, leaving a vacuum under it, and steam is introduced from the boiler above the piston. The piston therefore descends; and when it has arrived at the bottom of the cylinder, the position of the cock is changed to that represented in fig. 29. This change is made by turning the cock through one-fourth of an entire revolution, which may be done by a lever moved by the eccentric, or by various other means. One of the curved passages in the cock now opens a communication between the steam pipe and the bottom of the cylinder: while the other opens a communication between the condenser and the top of the cylinder. By these means, the steam from the boiler is introduced below the piston, while the steam above the piston is drawn off to the condenser. A vacuum being thus made above the piston, and steam introduced below it, ascends; and when it has arrived at the top of the cylinder, the cock being moved back, it resumes the position in fig. 28, and the same consequences ensue, the piston descends; and so the process is continued. In figs. 30, 31, the four-way cock with the passages to the top and bottom of the cylinder is represented on a larger scale.

This beautiful contrivance is not of late invention. It was used by Papin, and is also described by Leupold in his *Theatrum Machinarum*, a work published about the year 1720, in which an engine is described acting with steam of high pressure, on a principle which we shall describe in a subsequent chapter.

The four-way cock is liable to some practical objections. The quantity of steam which fills the tubes between the cock and the cylinder, is wasted every stroke. This objection, however, also applies to the sliding valve, (figs. 22, 23,) and to the sliding tube or D valves, (figs. 24, 25, 26, 27.) In fact, it is applicable to every contrivance in which means of shutting off the steam are not placed at both top and bottom of the cylinder. Besides this, however, the various passages and tubes cannot be conveniently made large enough to supply steam in sufficient abundance; and consequently it becomes necessary to produce steam in the boiler of a more than ordinary strength to bear the attenuation which it suffers in its passage through so many narrow tubes.

One of the greatest objections, however, to the use of the four-way cock, particularly in large engines, is its unequal wear. The parts of it near the passages having smaller surfaces, become more affected by the friction, and in a short

Pl. VII.

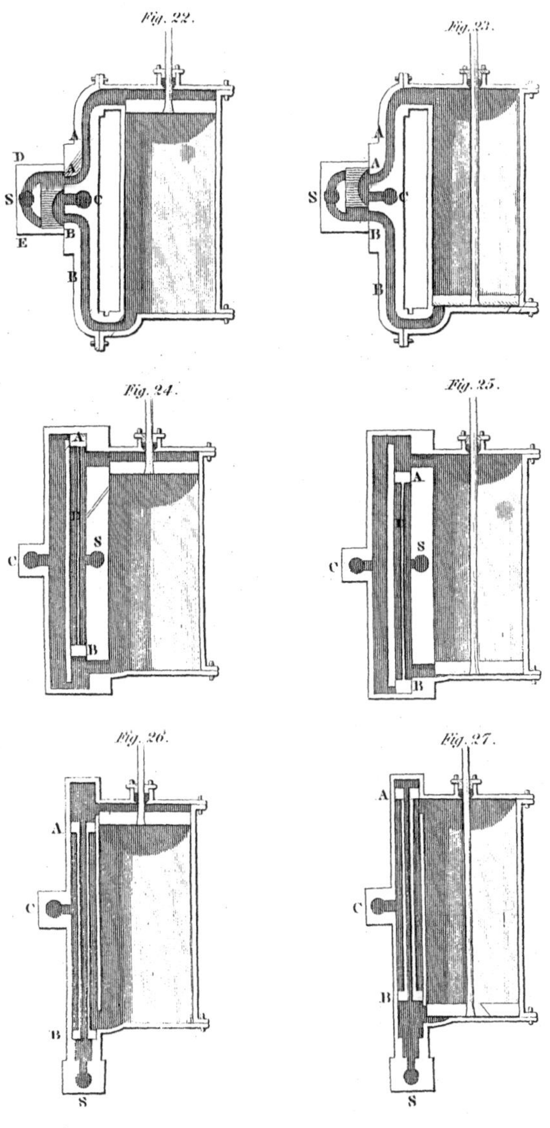


Drawn by the Author. Engr. by Peter Maverick.

time the steam leaks between the cock and its case, and becomes wasted, and tends to vitiate the vacuum. These cocks are seldom used in condensing engines, except they be small engines, but are frequently adopted in high pressure steam engines; for in these the leakage is not of so much consequence, as will appear hereafter.

---

# CHAPTER VIII.

## BOILER AND ITS APPENDAGES.—FURNACE.

The Boiler and its Appendages.—Level Gauges.—Feeding Apparatus.—Steam Gauge.—Barometer Gauge.—Safety Valves.—Self-regulating Damper.—Edelcrantz's Valve.—Furnace.—Self-consuming Furnace.—Brunton's self-regulating Furnace.—Oldham's Modification.

(65.) THE regular action of a steam engine, as well as the economy of fuel, depends in a great degree on the construction of the boiler or apparatus for generating the steam. The boiler may be conceived as a great magazine of steam for the use of the engine; and care must be taken not only that a sufficient quantity be always ready for the supply of the machine, but also that it shall be of the proper quality; that is, that its pressure shall not exceed that which is required, nor fall short of it. Precautions should, therefore, be taken that the production of steam should be exactly proportioned to the work to be done, and that the steam so produced shall be admitted to the cylinder in the same proportion.

To accomplish this, various contrivances, eminently remarkable for their ingenuity, have been resorted to, and which we shall now proceed to describe.

(*d*) It may be premised that boilers have been made of various figures, each having its own peculiar advantages and defects. That which possesses the greatest degree of strength is one of the shape of a cylinder. This form was originally introduced by Oliver Evans, in the United States. It will have its entire superiority when the fire is made beneath it in a furnace of masonry. But this method is not applicable to steamboats or locomotive engines. In these instances the weight of the separate furnace is an objection; when, therefore, this form is applied in them, the fire is usually made in a chamber of cylindric shape, within the boiler, and the smoke is conveyed to the chimney by flues, which pass through the water. These flues have, in some cases, been reduced to the size of small tubes, and of this method an example will be found in a subsequent part of this work. —A. E.

(66.) Different methods have been, from time to time, suggested for indicating the level of the water in the boiler. We have already mentioned the two guage pipes used in the earlier steam engines, (31); and which are still generally continued. There are, however, some other methods which merit our attention.

A weight F, (fig. 32,) half immersed in the water in the boiler, is supported by a wire, which, passing steam-tight through a small hole in the top, is connected by a flexible string or chain passing over a wheel W, with a counterpoise A, which is just sufficient to balance F, when half immersed. If F be raised above the water, A, being lighter, will no longer balance it, and F will descend, pulling up A, and turning the wheel W. If, on the other hand, F be plunged deeper in the water, A will more than balance it, and will pull it up. So that the only position in which F and A will balance each other is when F is half immersed. The wheel W is so adjusted, that when two pins, placed on its rim, are in the horizontal position, as in fig. 32, the water is at its proper level. Consequently it follows, that if the water rise above this level, the weight F is lifted, and A falls, so that the pins P P' come into the position in fig. 33. If, on the other hand, the level of the water falls, F falls and A rises, so that the pins P P assume the position in fig. 34. Thus, in general, the position of the pins P P′ becomes an indication of the quantity of water in the boiler.

Another method is to place a glass tube (fig. 35) with one end T entering the boiler above the proper level, and the other end T′ entering it below the proper level. It must be evident that the water in the tube will always stand at the same level as the water in the boiler; since the lower part has a free communication with that water, while the surface is submitted to the pressure of the same steam as the water in the boiler. This, and the last-mentioned gauge, have the advantage of addressing the eye of the engineer at once, without any adjustment; whereas the gauge cocks must be both opened, whenever the depth is to be ascertained.

These gauges, however, require the frequent attention of the engine-man; and it becomes desirable either to find some more effectual means of awakening that attention, or to render the supply of the boiler independent of any attention. In order to enforce the attention of the engine-man to replenish the boiler when partially exhausted by evaporation, a tube was sometimes inserted at the lowest level to which it was intended that the water should be permitted to fall. This tube was conducted from the boiler into the engine-house, where it terminated in a mouth-piece or whistle, so that, whenever the water fell below the level at which this tube was inserted in the boiler, the steam would rush through it, and, issuing with great velocity at the mouth-piece, would summon the engineer to his duty with a call that would rouse him even from sleep.

(67.) In the most effectual of these methods, the task of replenishing the boiler should still be executed by the engineer; and the utmost that the boiler itself was made to do was to give due notice of the necessity for the supply of water. The consequence was, among other inconveniences, that the level of the water was subject to constant variation.

To remedy this, a method has been invented by which the engine is made to feed its own boiler. The pipe G', (fig. 15,) which leads from the hot water pump H', terminates in a small cistern C, (fig. 36,) in which the water is received. In the bottom of this cistern a valve V is placed, which opens upward, and communicates with a feed-pipe, which descends into the boiler below the level of the water in it. The stem of the valve V is connected with a lever turning on the centre D, and loaded with a weight F, dipped in the water in the boiler in a manner similar to that described in fig. 32, and balanced by a counterpoise A in exactly the same way. When the level of the water in the boiler falls, the float F falls with it, and, pulling down the arm E of the lever, raises the valve V, and lets the water descend into the boiler from the cistern C. When the boiler has thus been replenished, and the level raised to its former place, F will again be raised, and the valve V closed by the weight A. In practice, however, the valve V adjusts itself by means of the effect of the water on the weight F, so as to permit the water from the feeding cistern C to flow in a continued stream, just sufficient in quantity to supply the consumption from evaporation, and to maintain the level of the water in the boiler constantly the same.

By this singularly felicitous arrangement, the boiler is made to replenish itself, or, more properly speaking, it is made to receive such a supply as that it never wants replenishing, an effect which no effort of attention on the part of an engine-man could produce. But this is not the only good effect produced by this contrivance. A part of the steam which originally left the boiler, and, having discharged its duty in moving the piston, was condensed and reconverted into water, and lodged by the air-pump in the hot-well, (47), is here again restored to the source from which it

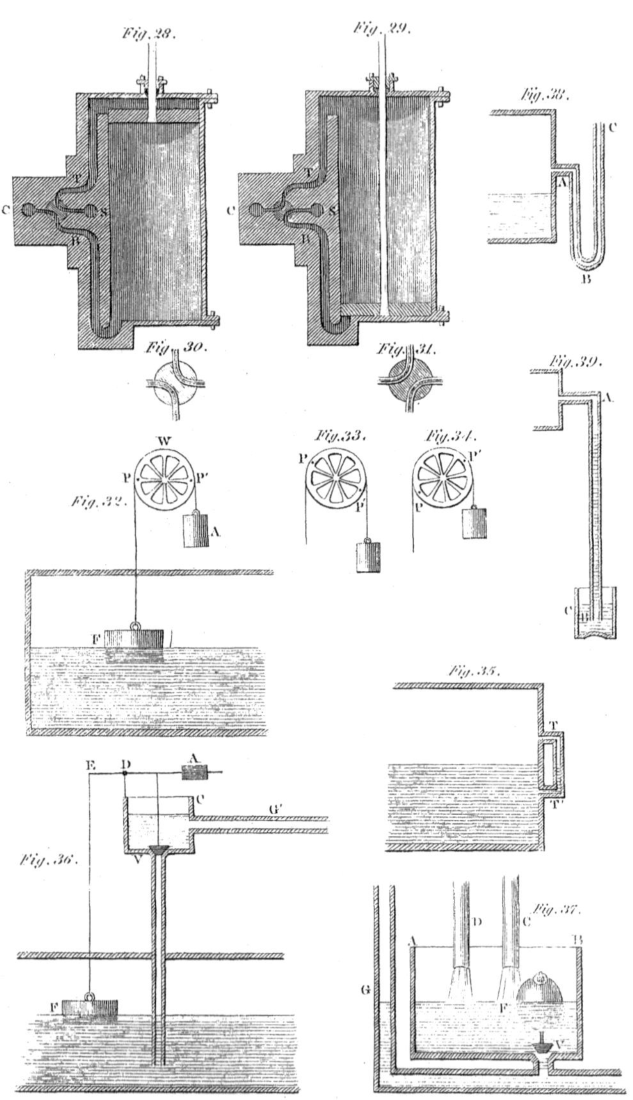
Fig. 28.
Fig. 29.
Fig. 30.
Fig. 31.
Fig. 32.
Fig. 33.
Fig. 34.
Fig. 35.
Fig. 36.
Fig. 37.
Fig. 38.
Fig. 39.

came, bringing back all the unconsumed portion of its heat preparatory to being once more put in circulation through the machine.

The entire quantity of hot water pumped into the cistern C is not always required for the boiler. A waste pipe may be provided for carrying off the surplus, which may be turned to any purpose for which it may be required; or it may be discharged into a cistern to cool, preparatory to being restored to the cold cistern, (fig. 12,) in case water for the supply of that cistern be not sufficiently abundant.

In cities and places in which it becomes an object to prevent the waste of water, the waste pipes proceeding from the feed cistern C, (fig. 36,) and from the cold cistern containing the condenser and air-pump, may be conducted to a cistern A B, (fig. 27.) Let C be the pipe from the feeding cistern, and D that from the cold cistern; by these pipes the waste water from both these cisterns is deposited in A B. In the bottom of A B is a valve V, opening upward, connected with a float F. When the quantity of water collected in the cistern A B is such that the level rises considerably, the float F is raised, and lifts the valve V, and the water flows into the main pipe, which supplies water for working the engine: G is the cold water pump for the supply of the cold cistern.

This arrangement for saving the water discharged from the feeding and condensing cisterns has been adopted in the printing office of the Bank of Ireland, and a very considerable waste of water is thereby prevented.

(68.) It is necessary to have a ready method of ascertaining at all times the strength of the steam which is used in working the engine. For this purpose a bent tube containing mercury is inserted into some part of the apparatus which has free communication with the steam. It is usually inserted in the *jacket* of the cylinder, (44.) Let A B C (fig. 38) be such a tube. The pressure of the steam forces the mercury down in the leg A B and up in the leg B C. If the mercury in both legs be at exactly the same level, the pres sure of the steam must be exactly equal to that of the atmosphere; because the steam pressure on the mercury in A B balances the atmospheric pressure on the mercury in B C. If, however, the level of the mercury in B C be above the level of the mercury in B A, the pressure of the steam will exceed that of the atmosphere. The excess of its pressure above that of the atmosphere may be found by observing the difference of the levels of the mercury in the tubes B C and B A; allowing a pressure of one pound on each square inch for every two inches in the difference of the levels.

If, on the contrary, the level of the mercury in B C should fall below its level in A B, the atmospheric pressure will exceed that of the steam, and the degree or quantity of the excess may be ascertained exactly in the same way.

If the tube be glass, the difference of levels of the mercury would be visible: but it is most commonly made of iron; and in order to ascertain the level, a thin wooden rod with a float is inserted in the open end of B C; so that the portion of the stick within the tube indicates the distance of the level of the mercury from its mouth. A bulb or cistern of mercury might be substituted for the leg A B, as in the common barometer. This instrument is called the *steam gauge.*

If the steam gauge be used as a measure of the strength of the steam which presses on the piston, it ought to be on the same side of the throttle valve (which is regulated by the governor) as the cylinder; for if it were on the same side of the throttle valve with the boiler, it would not be affected by the changes which the steam may undergo in passing through the throttle valve, when partially closed by the agency of the governor.

(69.) The force with which the piston is pressed depends on two things: 1°, the actual strength of the steam which presses on it; and 2°, on the actual strength of the vapour which resists it. For although the vacuum produced by the method of separate condensation be much more perfect than what had been produced in the atmospheric engines, yet still some vapour of a small degree of elasticity is found to be raised from the hot water in the bottom of the condenser before it can be extracted by the air-pump. One of these pressures is indicated by the steam gauge already described; but still, before we can estimate the force with which the piston descends, it is necessary to ascertain the force of the vapour which remains uncondensed, and resists the motion of the piston. Another gauge, called the barometer gauge, is provided for this purpose. A glass tube A B, (fig. 39,) more than thirty inches long, and open at both ends, is placed in an upright or vertical position, having the lower end B immersed in a cistern of mercury C. To the upper end is attached a metal tube, which communicates with the condenser, in which a constant vacuum, or rather high degree of rarefaction, is sustained. The same vacuum must, therefore, exist in the tube A B, above the level of the mercury; and the atmospheric pressure on the surface of the mercury in the cistern C will force the mercury up in the tube A B, until the column which is suspended in it is equal to the difference between the atmospheric pressure and the pressure of the uncondensed steam. The difference between the column of mercury sustained in this instrument and in the common barometer will determine the strength of the uncondensed steam, allowing a force proportional to one pound per square inch for every two inches of mercury in the difference of the two columns. In a well-constructed engine, which is in good order, there is very little difference between the altitude in the barometer gauge and the common barometer.

To compute the force with which the piston descends, thus becomes a very simple arithmetical process. First ascertain the difference of the levels of the mercury in the steam gauge. This gives the excess of the steam pressure above the atmospheric pressure. Then find the height of the mercury in the barometer gauge. This gives the excess of the atmospheric pressure above the uncondensed steam. Hence, if these two heights be added together, we shall obtain the excess of the impelling force of the steam from the boiler on the one side of the piston, above the resistance of the uncondensed steam on the other side. This will give the effective impelling force. Now, if one pound be allowed for every two inches of mercury in the two columns just mentioned, we shall have the number of pounds of impelling pressure on every square inch of the piston. Then if the number of square inches in the section of the piston be found, and multiplied by the number of pounds on each square inch, the whole effective force with which it moves will be obtained.

In the computation of the power of the engine, however, all this force, thus computed, is not to be allowed as the effective working power. For it requires some force, and by no means an inconsiderable portion, to move the engine itself, even when unloaded; all this, therefore, which is spent in overcoming friction, &c. is to be left out of account, and only the balance set down as the effective working power.

From what we have stated, it appears that in order to estimate the effective force with which the piston is urged, it is necessary to refer to both the barometer and the steam gauge. This double computation may be obviated by making one gauge serve both purposes. If the end C of the steam gauge, (fig. 38,) instead of communicating with the atmosphere, were continued to the condenser, we should have the pressure of the steam acting upon the mercury in the tube B A, and the pressure of the uncondensed vapour which resists the piston acting on the mercury in the tube B C. Hence the difference of the levels of the mercury in the tubes will at once indicate the difference between the force of the steam and that of the uncondensed vapour, which is the effective force with which the piston is urged.

(70.) To secure the boiler from accidents arising from the steam becoming too strong, a safety valve is used, similar to those described in Papin's steam engine, loaded with a weight equal to the strength which the steam is intended to have above the atmospheric pressure; for it is found expedient, even in condensing engines, to use the steam of a pressure somewhat above that of the atmosphere.

Besides this valve, another of the very opposite kind is sometimes used. Upon stopping the engine, and extinguishing the fire, it is found that the steam condensed within the boiler produces a vacuum; so that the atmosphere, pressing on the external surface of the boiler, has a tendency to crush it. To prevent this, a safety valve is provided, which opens inward, and which being forced open by the atmospheric pressure when a vacuum is produced within, the air rushes in, and a balance is obtained between the pressures within and without.

(71.) We have already explained the manner in which the governor regulates the supply of steam from the boiler to the cylinder, proportioning the quantity to the work to be done, and thereby sustaining a uniform motion. Since, then, the *consumption* of steam in the engine is subject to variation, owing to the various quantities of work it may have to perform, it is evident that the *production* of steam in the boiler should be subject to a proportional variation. For, otherwise, one of two effects would ensue: the boiler would either fail to supply the engine with steam, or steam would accumulate in the boiler, from being produced in too great abundance, and would escape at the safety valve, and thus be wasted. In order to vary the production of steam in proportion to the demands of the engine, it is necessary to increase or mitigate the furnace, as production is to be augmented or diminished. To effect this by any attention on the part of the engine-man would be impossible; but a most ingenious method has been contrived of making the boiler regulate itself in these respects. Let T (fig. 40) be a tube inserted in the top of the boiler, and descending nearly to the bottom

The pressure of the steam on the surface of the water in the boiler forces water up in the tube T, until the difference of the levels is equal to the difference between the pressure of the steam in the boiler and that of the atmosphere. A weight F, half immersed in the water in the tube, is suspended by a chain which passes over the wheels P P', and is balanced by a metal plate D, in the same manner as the float in fig. 32 is balanced by the weight A. The plate D passes through the mouth of the flue E, as it issues finally from the boiler; so that when the plate D falls, it stops the flue, and thereby suspends the draught of air through the furnace, mitigates the fire, and diminishes the production of steam. If, on the contrary, the plate D be drawn up, the draught is increased, the fire rendered more effective, and the production of steam in the boiler stimulated. Now, suppose that the boiler is producing steam faster than the engine consumes it, either because the load on the engine has been diminished, and therefore its consumption of steam, proportionally diminished, or because the fire has become too intense. The consequence is, that the steam, beginning to accumulate, will press upon the surface of the water in the boiler with increased force, and the water will rise in the tube T. The weight E will therefore be lifted, and the plate D will descend, stop the draught, mitigate the fire, and check the production of steam; and it will continue so to do until the production of steam becomes exactly equal to the demands of the engine.

If, on the other hand, the production of steam be not equal to the wants of the machine, either because of the increased load, or the insufficiency of the fire, the steam in the boiler losing its elasticity, the surface of the water rises, not sustaining a pressure sufficient to keep it at its wonted level. Therefore, the surface in the tube T falls, and the weight F falls, and the plate D rises. The draught is thus increased, by opening the flue, and the fire rendered more intense; and thus the production of steam is stimulated, until it is sufficiently rapid for the purposes of the engine. This apparatus is called the *self-acting damper.*

(72.) It has been proposed to connect this damper with the safety valve invented by the Chevalier Edelcrantz. A small brass cylinder is fixed to the boiler, and is fitted with a piston which moves in it, without much friction, and nearly steam-tight. The cylinder is closed at top, having a hole through which the piston-rod plays; so that the piston is thus prevented from being blown out of the cylinder by the steam. The side of the cylinder is pierced with small holes opening into the air, and placed at short distances above each other. Let the piston be loaded with a weight proportional to the pressure of the steam intended to be produced. When the steam has acquired a sufficient elasticity, the piston will be lifted, and steam will escape through the first hole. If the production of steam be not too rapid, and that its pressure be not increasing, the piston will remain suspended in this manner: but if it increase, the piston will be raised above the second hole, and it will continue to rise until the escape of the steam through the holes is sufficient to render the weight of the piston a counterpoise for the steam. This safety valve is particularly well adapted to cases where steam of an exactly uniform pressure is required; for the pressure must necessarily be always equal to the weight on the piston. Thus, suppose the section of the piston be equal to a square inch; if it be loaded with 10lbs., including its own weight, the steam which will sustain it in any position in the cylinder, whether near the bottom or top, must always be exactly equal in pressure to 10lbs. per inch. In this respect it resembles the quality already explained in the governor, and renders the pressure of the steam uniform, exactly in the same manner as the governor renders the velocity of the engine uniform.

(73.) The economy of fuel depends, in a great degree, on the construction of the furnace, independently of the effects of the arrangements we have already described.

The grate or fireplace of an ordinary furnace is placed under the boiler; and the atmospheric air, passing through the ignited fuel, supplies sufficient oxygen to support a large volume of flame, which is carried by the draught into a flue, which circulates twice or oftener round the boiler, and in immediate contact with it, and finally issues into the chimney. Through this flue the flame circulates, so as to act on every part of the boiler near which the flue passes; and it is frequently not until it passes into the chimney, and sometimes not until it leaves the chimney, that it ceases to exist in the state of flame.

The dense black smoke which is observed to issue from the chimneys of furnaces is formed of a quantity of unconsumed fuel, and may be therefore considered as so much fuel wasted. Besides this, in large manufacturing towns, where a great number of furnaces are employed, it is found that the quantity of smoke which thus becomes diffused through the atmosphere renders it pernicious to the health and destructive to the comforts of the inhabitants.

These circumstances have directed the attention of engineers to the discovery of means whereby this smoke or wasted fuel may be consumed for the use of the engine itself, or for whatever use the furnace may be applied to. The most usual method of accomplishing this is by so arranging matters that fuel in a state of high combustion, and, therefore, producing no smoke, shall be always kept on that part of the grate which is nearest to the mouth of the flue, (and which we shall call the back;) by this means the smoke which arises from the imperfectly ignited fuel which is nearer to the front of the grate must pass over the surface of the red fuel, before it enters the flue, and is thereby ignited, and passes in a state of flame into the flue. A passage called the *feeding-mouth* leads to the front of the grate, and both this passage and the grate are generally inclined at a small angle to the horizon, in order to facilitate the advance of the fuel according as its combustion proceeds.

When fresh fuel for feeding the boiler is first introduced, it is merely laid in the feeding mouth. Here it is exposed to the action of a part of the heat of the burning fuel on the grate, and undergoes, in some degree, the process of coking. The door of the feeding mouth is furnished with small apertures for the admission of a stream of air, which carries the smoke evolved by the coking of the fresh fuel over the burning fuel on the grate, by which this smoke is ignited, and becomes flame, and in this state enters the flue, and circulates round the boiler. When the furnace is to be fed, the door of the feeding mouth is opened, and the fuel which had been laid in it, and partially coked, is forced upon the front part of the grate. At first, its combustion being imperfect, but proceeding rapidly, a dense black smoke arises from i.. The current of air from the open door through the feeding mouth carries this over the vividly burning fuel in the back part of the grate, by which the smoke, being ignited, passes in a state of flame into the flue. When the furnace again requires feeding, every part of this fuel will be in a state of active combustion, and it is forced to the back part of the grate next the flue, preparatory to the introduction of more fuel from the feeding mouth.

The apertures in the door of the feeding mouth are furnished with covers, so that the quantity of air admitted through them can be regulated by the workmen. The efficiency of these furnaces in a great degree depends on the judicious admission of the air through the feeding mouth; for if less than the quantity necessary to support the combustion of the fuel be admitted, a part of the smoke will remain unconsumed; and if more than the proper quantity be admitted, it will defeat the effects of the fuel by cooling the boiler. If the process which we have just described be considered, it will not be difficult to perceive the total impossibility in such a furnace of exactly regulating the draught of air, so that too much shall not pass at one time, and too little at another. When the door is open to introduce fresh fuel into the feeding mouth, and advance that which occupied it upon the grate, the workman ceases to have any control whatever over the draught of air; and even at other times when the door is closed, his discretion and attention cannot be depended on. The consequence is, that with these defects the proprietors of steam engines found, that in the place of economizing the fuel, the use of these furnaces entailed on them such an increased expense that they were generally obliged to lay them aside.

(74.) Mr. Brunton of Birmingham, having turned his attention to the subject, has produced a furnace which seems to be free from the objections against those we have just mentioned. The advantages of his contrivance, as stated by himself, are as follow: —

"First, I put the coal upon the grate by small quantities, and at very short intervals, say every two or three seconds. 2dly, I so dispose of the coals upon the grate, that the smoke evolved must pass over that part of the grate upon which the coal is in full combustion, and is thereby consumed. 3dly, As the introduction of coal is uniform in short spaces of time, the introduction of air is also uniform, and requires no attention from the fireman.

"As it respects economy: 1st, The coal is put upon the fire by an apparatus driven by the engine, and so contrived that the quantity of coal is proportioned to the quantity of work which the engine is performing, and the quantity of air admitted to consume the smoke is regulated in the same manner. 2dly, The fire door is never opened, excepting to clean the fire; the boiler of course is not exposed to that continual irregularity of temperature which is unavoidable in the common furnace, and which is found exceedingly injurious to boilers. 3dly, The only attention required is to fill the coal receiver, every two or three hours, and clean the fire when necessary. 4thly, The coal is more completely consumed than by the common furnace, as all the effect of what is termed stirring up the fire (by which no inconsider able quantity of coal is passed into the ash-pit) is attained without moving the coal upon the grate."

The fireplace is a circular grate placed on a vertical shaft in a horizontal position. It is capable of revolving, and is made to do so by the vertical shaft, which is turned by wheel-work, which is worked by the engine itself; or this shaft or spindle may be turned by a water-wheel, on which a stream of water is allowed to flow from a reservoir into which it is pumped by the power of the engine; and by regulating the quantity of water in the stream, the grate may be made to revolve with a greater or less speed. In that part of the boiler which is over the grate, there is an aperture, in which is placed a hopper, through which fuel is let down upon the grate at the rate of any quantity per minute that may be required. The apparatus which admits the coals through this hopper is worked by the engine also, and by the same means as the grate is turned, so that the grate revolves with a speed proportional to the rapidity with which the fuel is admitted through the hopper, and by this ingenious arrangement the fuel falls equally thick upon the grate.

The supply of water which turns the wheel which works the grate and the machinery in the hopper, is regulated by a cock connected with the self-regulating damper; so that when the steam is being produced too fast, the supply of water will be diminished, and by that means the supply of fuel to the grate will be diminished, and the grate will revolve less rapidly: and when the steam is being produced too slowly for the demands of the engine, the contrary effects take place. In this way the fuel which is introduced into the furnace is exactly proportioned to the work which the engine has to perform. The hopper may be made large enough to hold coals for a day's work, so that the furnace requires no other attendance than to deposite coals in the hopper each morning.

The coals are let down from the hopper on the grate at that part which is most remote from the flue; and as they descend in very small quantities at a time, they are almost immediately ignited. But until their ignition is complete, a smoke will arise, which, passing to the flue over the vividly burning fuel, will be ignited. Air is admitted through proper apertures, and its quantity regulated by the damper in the same manner as the supply of fuel.

The superiority of this beautiful invention over the common smoke-consuming furnaces is very striking. Its principle of self-adjustment as to the supply of coals and atmospheric air, and the proportioning of these to the quantity of work to be performed by the engine, not only independent of human labour, but with a greater degree of accuracy than any human skill or attention could possibly effect, produces saving of expense, both in fuel and labour.

(75.) Mr. Oldham, engineer to the Bank of Ireland, has proposed another modification of the self-regulating furnace, which seems to possess several advantages, and evinces considerable ingenuity.

He uses a slightly inclined grate, at the back or lower end of which is the flue, and at the front or higher end, the hopper for admitting the coals. In the bottom or narrow end of the hopper is a moveable shelf, worked by the engine. Upon drawing back this shelf, a small quantity of fuel is allowed to descend upon a fixed shelf under it; and upon the return of the moveable shelf, this fuel is protruded forward upon the grate. Every alternate bar of the grate is fixed, but the intermediate ones are connected with levers, by which they are moved alternately up and down.* The effect is, that the coals upon the bars are continually stirred, and gradually advanced by their own weight from the front of the grate, where they fall from the hopper, to the back, where they

* Mr. Brunton used moveable bars in a furnace constructed by him before he adopted the horizontal revolving grate. That plan, however, does not appear to have been as successful as the latter, as he has abandoned it. Mr. Oldham states that his furnace has been in use for several years without any appearance of derangement in the mechanism, and with a considerable saving of fuel.

are deposited in the ash-pit. By the shape and construction of the bars, the air is conducted upward between them, and rushes through the burning fuel, so as to act in the manner of a blowpipe, and the entire surface of the fire presents a sheet of flame.

We cannot fail to be struck with the beauty of all these contrivances, by which the engine is made to regulate itself, and supply its own wants. It is, in fact, all but alive. It was observed by Belidor, long before the steam engine reached the perfection which it has now acquired, that it resembled an animal, and that no mere work of man ever approached so near to actual *life*. Heat is the principle of its existence. The boiler acts the part of the heart, from which its vivifying fluid rushes copiously through all the tubes, where, having discharged the various functions of life, and deposited its heat in the proper places, it returns again to the source it sprung from, to be duly prepared for another circulation. The healthfulness of its action is indicated by the regularity of its pulsations; it procures its own food by its own labour; it selects those parts which are fit for its support, both as to quantity and quality; and has its natural evacuations, by which all the useless and innutritious parts are discharged. It frequently cures its own diseases, and corrects the irregularity of its own actions, exerting something like moral faculties. Without designing to carry on the analogy, Mr. Farcy, in speaking of the variations incident to the work performed by different steam engines, states some further particulars in which it may be curiously extended. "We must observe," says he, "that the variation in the performance of different steam engines, which are constructed on the same principle, working under the same advantages, is the same as would be found in the produce of the labour of so many different horses or other animals, when compared with their consumption of food; for the effects of different steam engines will vary as much from small differences in the proportion of their parts, as the strength of animals from the vigour of their constitutions; and again, there will be as great differences in the performance of the same engine when in good and bad order, from all the parts being tight and well oiled, so as to move with little friction, as there is in the labour of an animal from his being in good or bad health, or excessively fatigued: but in all these cases there will be a maximum which cannot be exceeded, and an average which we ought to expect to obtain.

---

## CHAPTER IX.

### DOUBLE-CYLINDER ENGINES.

Hornblower's Engine.—Woolf's Engine.—Cartwright's Engine.

(76.) THE expansive property of steam, of which Watt availed himself in his single engine by cutting off the supply of steam before the descent of the piston was completed, was applied in a peculiar manner by an engineer named Hornblower, about the year 1781, and at a later period by Woolf. Hornblower was the first who conceived the idea of working an engine with two cylinders of different sizes, by allowing the steam to flow freely from the boiler until it fills the smaller cylinder, and then permitting it to expand into the greater one, employing it thus to press down two pistons in the manner which we shall presently describe. The condensing apparatus of Hornblower, as well as the other appendages of the engine, do not differ materially from those of Watt; so that it will be sufficient for our present purpose to explain the manner in which the steam is made to act in moving the piston.

Let C, fig. 41, be the centre of the great working beam, carrying two arch heads, on which the chains of the piston-rods play. The distances of these arch heads from the centre

C must be in the same proportion as the length of the cylinders, in order that the same play of the beam may correspond to the plays of both pistons. Let F be the steam pipe from the boiler, and G a valve to admit the steam above the lesser piston. H is a tube by which a communication may be opened by the valve I between the top and bottom of the lesser cylinder B. K is a tube communicating, by the valve L, between the bottom of the lesser cylinder B and the top of the greater cylinder A. M is a tube communicating, by the valve N, between the top and bottom of the greater cylinder A; and P a tube leading to the condenser by the exhausting valve O.

At the commencement of the operation, suppose all the valves opened, and steam allowed to flow through the entire engine until the air be completely expelled, and then let all the valves be closed. To start the engine, let the exhausting valve O and the steam valves G and L be opened, as in fig. 41. The steam will flow freely from the boiler, and press upon the lesser piston, and at the same time the steam below the greater piston will flow into the condenser, leaving a vacuum in the greater cylinder. The valve L being opened, the steam which is under the piston in the lesser cylinder will flow through K, and press on the greater piston, which, having a vacuum beneath it, will consequently descend. At the commencement of the motion, the lesser piston is as much resisted by the steam below it as it is urged by the steam above it; but after a part of the descent has been effected, the steam below the lesser piston, passing into the greater, expands into an increased space, and therefore loses its elastic force proportionally. The steam above the lesser piston retaining its full force by having a free communication with the boiler by the valve G, the lesser piston will be urged by a force equal to the excess of the pressure of this steam above the diminished pressure of the expanded steam below it. As the pistons descend, the steam which is between it continually increasing in its bulk, and therefore decreasing in its pressure, from whence it follows, that the force which resists the lesser piston is continually decreasing, while that which presses it down remains the same, and, therefore, the effective force which impels it must be continually increasing.

On the other hand, the force which urges the greater piston is continually decreasing, since there is a vacuum below it, and the steam which presses it is continually expanding into an increased bulk.

Impelled in this way, let us suppose the pistons to have arrived at the bottoms of the cylinders, as in fig. 42, and let the valves G, L, and O be closed, and the valves I and N opened. No steam is allowed to flow from the boiler, G being closed, nor any allowed to pass into the condenser, since O is closed, and all communications between the cylinders is stopped by closing L. By opening the valve I, a free comnication is made between the top and bottom of the lesser piston through the tube H, so that the steam which presses above the lesser piston will exert the same pressure below it, and the piston is in a state of indifference. In the same manner the valve N being open, a free communication is made between the top and bottom of the greater piston, and the steam circulates above and below the piston, and leaves it free to rise. A counterpoise attached to the pump rods in this case draws up the piston, as in Watt's single engine; and when they arrive at the top, the valves I and N are closed, and G, L, and O opened, and the next descent of the piston is produced in the manner already described, and so the process is continued.

The valves are worked by the engine itself, by means similar to some of those already described. By computation, we find the power of this engine tc be nearly the same as a similar engine on Watt's expansive principle. It does not, however, appear that any adequate advantage was gained by this modification of the principle, since no engines of this construction are now made.

(77.) The use of two cylinders was revived by *Arthur Woolf*, in 1804, who, in this and the succeeding year, obtained patents for the application of steam raised under a high pressure to double-cylinder engines. The specification of his patent states, that he has proved by experiment that steam raised under a safety valve loaded with any given number of pounds upon the square inch, will, if allowed to expand into as many times its bulk as there are pounds of pressure on the square inch, have a pressure equal to that of the atmosphere. Thus, if the safety valve be loaded with four pounds on the square inch, the steam, after expanding into four times its bulk, will have the atmospheric pressure If it be loaded with 5, 6, or 10 lbs. on the square inch, it will have the atmospheric pressure when it has expanded into 5, 6, or 10 times its bulk, and so on. It was, however, understood in this case, that the vessel into which it was allowed to expand should have the same temperature as the steam before it expands.

It is very unaccountable how a person of Mr. Woolf's experience in the practical application of steam could be led into errors so gross as those involved in the averments of this patent; and it is still more unaccountable how the experiments could have been conducted which led him to conclusions not only incompatible with all the established properties of elastic fluids—properties which at that time were perfectly understood—but even involving in themselves palpable contradiction and absurdity. If it were admitted that every additional pound avoirdupois which should be placed upon the safety valve would enable steam, by its expansion into a proportionally enlarged space, to attain a pressure equal to the atmosphere, the obvious consequence would be, that a physical relation would subsist between the atmospheric pressure and the pound avoirdupois! It is wonderful that it did not occur to Mr. Woolf, that, granting his principle to be true at any given place, it would necessarily be false at another place where the barometer would stand at a different height. Thus if the principle were true at the foot of a mountain, it would be false at the top of it; and if it were true in fair weather, it would be false in foul weather, since these circumstances would be attended by a change in the atmospheric pressure, without making any change in the pound avoirdupois.*

The method by which Mr. Woolf proposed to apply the principle which he imagined himself to have discovered was by an arrangement of cylinders similar to those of Hornblower, but having their magnitudes proportioned to the greater extent of expansion which he proposed to use. Two cylinders, like those of Hornblower, were placed under the working beam, having their piston-rods at distances from the axis proportioned to the lengths of their respective strokes. The relative magnitudes of the cylinders A and B must be adjusted according to the extent to which the principle of expansion is intended to be used. The valves C C′ were placed at each end of the lesser cylinder in tubes communicating with the boiler, so as to admit steam on each side of the lesser piston, and cut it off at pleasure. A tube, D′, formed a communication between the upper end of the lesser and lower end of the greater cylinder, which communication is opened and closed at pleasure by the valve E′. In like manner, the tube D forms a communication between the lower end of the lesser cylinder and the upper end of the greater, which may be opened and closed by the valve E. The top and bottom of the greater cylinder communicated with the condenser by valves F′ F.

Let us suppose that the air is blown from the engine in the usual way, all the valves closed, and the engine ready to

* It is strange that this absurdity has been repeatedly given as unquestionable fact in various encyclopædias on the article "Steam Engine," as well as in by far the greater number of treatises expressly on the subject.

start, the pistons being at the top of the cylinders. Open the valves C, E, and F. The steam which occupies the greater cylinder below the piston will now pass into the condenser through F, leaving a vacuum below the piston. The steam which is in the lesser cylinder below the piston will pass through D and open the valve E, and will press down the greater piston. The steam from the boiler will flow in at C, and press on the lesser piston. At first the whole motion will proceed from the pressure upon the greater piston, since the steam, both above and below the lesser piston, has the same pressure. But, as the pistons descend, the steam below the less passing into the greater cylinder, expands into a greater space, and consequently exerts a diminished pressure, and, therefore, the steam on the other side exerting an undiminished pressure, acquires an impelling force exactly equal to the pressure lost in the expansion of the steam between the two pistons. Thus both pistons will be pressed to the bottoms of their respective cylinders. It will be observed that in the descent the greater piston is urged by a continually decreasing force, while the lesser is urged by continually increasing force.

Upon the arrival of the pistons at the bottoms of the cylinders, let the valves, C, E, F be closed, and C′, E′, F′ be opened, as in fig. 44. The steam which is above the greater piston now flows through F′ into the condenser, leaving the space above the piston a vacuum. The steam which is above the lesser piston passes through E′ and D′ below the greater, while the steam from the boiler is admitted through C′ below the lesser piston. The pressure of the steam entering through E′ below the greater piston, pressing on it against the vacuum above it, commences the ascent. In the mean time the steam above the lesser piston passing into the enlarged space of the greater cylinder, loses gradually its elastic force, so that the steam entering from the boiler at C′ becomes in part effective, and the ascent is completed under exactly the same circumstances as a descent, and in this way the process is continued.

It is evident that the valves may be easily worked by the mechanism of the engine itself.

In this arrangement the pistons ascend and descend together, and their rods must consequently be attached to the beam at the same side of the centre. It is sometimes desirable that they should act on different sides of the centre of the beam, and consequently that one should ascend while the other descends. It is easy to arrange the valves so as to effect this. In fig. 45, the lesser piston is at the bottom of the cylinder, and the greater at the top. On opening the valves C', E' F', a vacuum is produced below the greater piston, and steam flows from the lesser cylinder, through E', *above* the greater piston, and presses it down. At the same time steam being admitted from the boiler through C' below the lesser piston, forces it up against the diminishing force of the steam above it, which expands into the greater cylinder. Thus as the greater piston descends the lesser ascends. When each has traversed its cylinder, the valves C', E', F' being closed, and C, E, F opened, the lesser piston will descend, and the greater ascend, and so on.

(78.) The law according to which the elastic force of steam diminishes as it expands, of which Mr. Woolf appears to have been entirely ignorant, is precisely similar to the same property in air and other elastic fluids. If steam expands into twice or thrice its volume, it will lose its elastic force in precisely the same proportion as it enlarges its bulk; and therefore will have only a half or a third of its former pressure, supposing that as it expands its temperature is kept up. Although Mr. Woolf's patent contained the erroneous principle which we have noticed, yet, so far as his invention suggested the idea of employing steam at a very high pressure, and allowing it to expand in a much greater degree than was contemplated either by Watt or Hornblower, it became the means of effecting a considerable saving in fuel;

for engines used for pumping on a large scale, the steam being produced under a pressure of forty or fifty pounds or more upon the square inch, might be worked first through a small space with intense force, and the communication with the boiler being then cut off, it might be allowed, with great advantage, to expand through a very large space. Some double-cylinder engines upon this principle have been worked in Cornwall, with considerable economy. But the form in which the expansive principle, combined with high pressure, is now applied in the engines used for raising water from the mines, is that in which it was originally proposed by Watt. A single cylinder of considerable length is employed; the piston is driven through a small proportion of this length by steam, admitted from the boiler at a very intense pressure; the steam being then cut off, the piston is urged by the expansive force of the steam which has been admitted, and is by that means brought to the bottom of the cylinder.

It is evident, under such circumstances, that the pressure of the steam admitted from the boiler must be much greater than the resistance opposed to the piston, and that the motion of the piston must, in the first instance, be accelerated and not uniform. If the piston moved from the commencement with a uniform motion, the pressure of the steam urging it must necessarily be exactly equal to the resistance opposed to it, and then cutting off the supply of steam from the boiler, the piston could only continue its motion by inertia, the steam immediately becoming of less pressure than the resistance; and after advancing through a very small space, the piston would recoil upon the steam, and come to a state of rest. The steam, however, at the moment it is cut off, being of much greater pressure than the amount of resistance upon the piston, will continue to drive the piston forward, until, by its expansion, its force is so far diminished as to become equal to the resistance of the piston. From that point the impelling power of the steam will cease, and the piston will move forward by its inertia only. The point at which the steam is cut off should therefore be so regulated that it shall acquire a pressure equal to the resistance on the piston by its expansion, just at such a distance from the end of the stroke as the piston may be able to move through by its inertia. It is evident the adjustment of this will require great care and nicety of management.

(79.) In 1797 a patent was granted to the *Rev. Mr. Cartwright*, a gentleman well known for other mechanical inventions, for some improvements in the steam engine. His contrivance is at once so elegant and simple, that, although it has not been carried into practice, we cannot here pass it over without notice.

The steam pipe from the boiler is represented cut off at B, (fig. 46;) T is a spindle valve for admitting steam above the piston, and R is a spindle valve in the piston; D is a curved pipe forming a communication between the cylinder and the condenser, which is of very peculiar construction. Cartwright proposed effecting a condensation without a jet, by exposing the steam to contact with a very large quantity of cold surface. For this purpose, he formed his condenser by placing two cylinders nearly equal in size one within the other, allowing the water of the cold cistern in which they were placed to flow through the inner cylinder, and to surround the outer one. Thus the thin space between the two cylinders formed the condenser.

The air-pump is placed immediately under the cylinder, and the continuation of the piston-rod works its piston, which is solid and without a valve: F is the pipe from the condenser to the air-pump, through which the condensed steam is drawn off through the valve G on the ascent of the piston, and on the descent, this is forced through a tube into a hot well H, for the purpose of feeding the boiler through the feed pipe I. In the top of the hot well H is a valve which opens inwards, and is kept closed by a ball floating on the surface of the liquid. The pressure of the condensed air

Pl. IX.

Fig. 40.
P
P′
T
D
Fig. 41.
C
F
G
H
B
K
A
M
N
I
L
P
O
Fig. 42.
F
G
H
B
K
A
M
N
I
L
P
O
Fig. 43.
E′
E
C
F′
B
D
A
C′
D′
F
Fig. 44.
E′
E
C
F
D
C′
D
F
Fig. 45.
1
2
C
E′
F
C′
E
F′

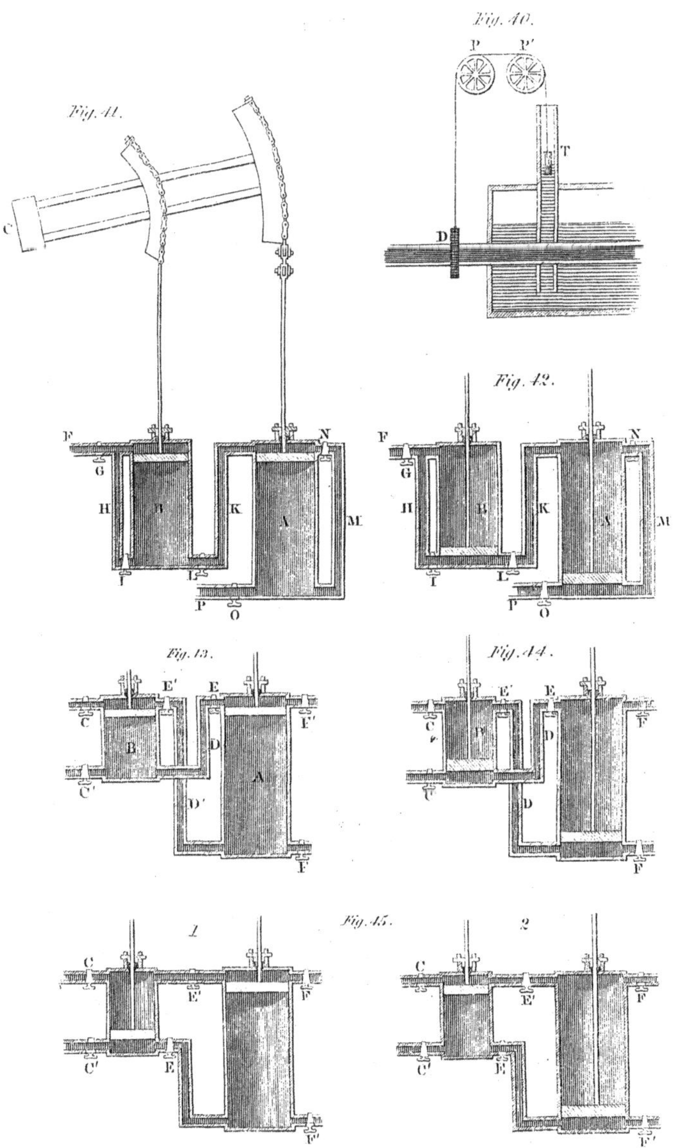

Pl. X.

*Fig. 46.*

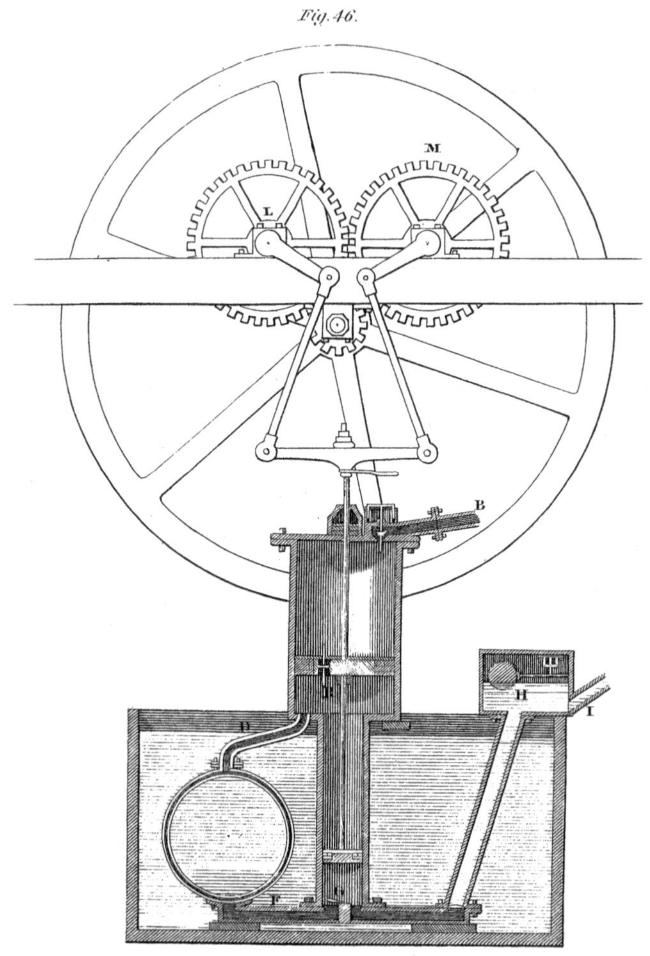


*Engr. by P. Maverick.*

above the surface of the liquid in H forces it through I into the boiler. When the air accumulates in too great a degree in H, the surface of the liquid is pressed so low that the ball falls and opens the valve, and allows it to escape. The air in H is that which is pumped from the condenser with the liquid, and which was disengaged from it.

Let us suppose the piston at the top of the cylinder; it strikes the tail of the valve T, and raises it, while the stem of the piston valve R strikes the top of the cylinder, and is pressed into its seat. A free communication is at the same time open between the cylinder, below the piston and the condenser, through the tube D. The pressure of the steam thus admitted above the piston, acting against the vacuum below it, will cause its descent. On arriving at the bottom of the cylinder, the tail of the piston valve R will strike the bottom, and it will be lifted from its seat, so that a communication will be opened through it with the condenser. At the same moment, a projecting spring K, attached to the piston-rod, strikes the stem of the steam valve T, and presses it into its seat. Thus, while the further admission of steam is cut off, the steam above the piston flows into the condenser, and the piston, being relieved from all pressure, is drawn up by the momentum of the fly wheel, which continues the motion it received from the descending force. On the arrival of the piston again at the top of the cylinder, the valve T is opened and R closed, and the piston descends as before, and so the process is continued.

The mechanism by which motion is communicated from the piston to the fly wheel is peculiarly elegant. On the axis of the fly wheel is a small wheel with teeth, which work in the teeth of another large wheel L. This wheel is turned by a crank, which is worked by a cross piece attached to the end of the piston-rod. Another equal-toothed wheel M is turned by a crank, which is worked by the other end of the cross arm attached to the piston-rod.

One of the peculiarities of this engine is, that the liquid which is used for the production of steam in the boiler circulates through the machine without either diminution or admixture with any other fluid, so that the boiler never wants more feeding than what can be supplied from the hot-well H. This circumstance forms a most important feature in the machine, as it allows of ardent spirits being used in the boiler, instead of water, which, since they boil at low heats, promised a saving of half the fuel. The inventor even proposed that the engine should be used as a still, as well as a mechanical power, in which case the whole of the fuel would be saved.

In this engine, the ordinary method of rendering the piston steam-tight, by oil or melted wax or tallow poured upon it, could not be applied, since the steam above the piston must always have a free passage through the piston valve R. The ingenious inventor therefore contrived a method of making the piston steam-tight in the cylinder, without oil or stuffing, and his method has since been adopted with success in other engines.

A ring of metal is ground into the cylinder, so as to fit it perfectly, and is then cut into four equal segments. The inner surface of this ring being slightly conical, another ring is ground into it, so as to fit it perfectly, and this is also cut into four segments, and one is placed within the other, but in such a manner that the joints or divisions do not coincide. The arrangement of the two rings is represented in fig. 47. Within the inner ring are placed four springs, which press the pieces outward against the sides of the cylinder, and are represented in the diagram. Four pairs of these rings are placed one over another, so that their joints do not coincide, and the whole is screwed together by plates placed at top and bottom. A vertical section of the piston is given in fig. 48.

One of the advantages of this piston is, that the longer it

is worked, the more accurately it fits the cylinder, so that as the machine wears it improves.

Metallic pistons have lately come into very general use, and such contrivances differ very little from the above.

---

## CHAPTER X

### LOCOMOTIVE ENGINES ON RAILWAYS.

High-pressure Engines.—Leupold's Engine.—Trevithick and Vivian.—Effects of Improvement in Locomotion.—Historical Account of the Locomotive Engine.—Blenkinsop's Patent.—Chapman's Improvement.—Walking Engine.—Stephenson's First Engines.—His Improvements.—Liverpool and Manchester Railway Company.—Their preliminary Proceedings.—The great Competition of 1829.—The Rocket.—The Sanspareil.—The Novelty.—Qualities of the Rocket.—Successive Improvements.—Experiments.—Defects of the present Engines.—Inclined Planes.—Methods of surmounting them.—Circumstances of the Manchester Railway Company.—Probable Improvements in Locomotives.—Their Capabilities with respect to Speed.—Probable Effects of the projected Railroads.—Steam Power compared with Horse Power.—Railroads compared with Canals.

(80.) In the various modifications of the steam engine which we have hitherto considered, the pressure introduced on one side of the piston derives its efficacy either wholly or partially from the vacuum produced by condensation on the other side. This always requires a condensing apparatus, and a constant and abundant supply of cold water. An engine of this kind must therefore necessarily have considerable dimensions and weight, and is inapplicable to uses in which a small and light machine only is admissible. If the condensing apparatus be dispensed with, the piston will always be resisted by a force equal to the atmospheric pressure, and the only part of the steam pressure which will be available as a moving power, is that part by which it ex ceeds the pressure of the atmosphere. Hence, in engines which do not work by condensation, steam of a much higher pressure than that of the atmosphere is indispensably necessary, and such engines are therefore called *high-pressure engines.*

We are not, however, to understand that every engine, in which steam is used of a pressure exceeding that of the atmosphere, is what is meant by a *high-pressure engine;* for in the ordinary engines in common use, constructed on Watt's principle, the safety valve is loaded with from 3 to 5 lbs. on the square inch; and in Woolf's engines, the steam is produced under a pressure of 40 lbs. on the square inch. These would therefore be more properly called *condensing engines* than *high-pressure engines;* a term quite inapplicable to those of Woolf. In fact, by *high-pressure engines* is meant engines in which no vacuum is produced, and, therefore, in which the piston works against a pressure equal to that of the atmosphere.

In these engines, the whole of the condensing apparatus, *viz.* the cold-water cistern, condenser, air-pump, cold-water pump, &c. are dispensed with, and nothing is retained except the boiler, cylinder, piston, and valves. Consequently, such an engine is small, light, and cheap. It is portable also, and may be moved, if necessary, along with its load, and is therefore well adapted to locomotive purposes.

(81.) High-pressure engines were one of the earliest modifications of the steam engine. The contrivance, which is obscurely described in the article already quoted, (27), from the Century of Inventions, is a high-pressure engine; for the power there alluded to is the elastic force of steam working against the atmospheric pressure. *Newcomen,* in 1705, applied the working beam, cylinder, and piston to the atmospheric engine; and *Leupold,* about 1720, combined the working beam and cylinder with the high-pressure principle, and produced the earliest high-pressure engine worked by a cylinder and piston. The following is a description of *Leupold's* engine: —

A (fig. 49) is the boiler, with the furnace beneath it; C C' are two cylinders with two solid pistons, P P', connected with the working beams B B', to which are attached the pump rods, R R', of two forcing pumps, F F', which communicate with a great force pipe S; G is a *four-way cock*, (66), already described. In the position in which it stands in the figure, the steam is issuing from below the piston P into the atmosphere, and the piston is descending by its own weight; steam from the boiler is at the same time pressing up the piston P', with a force equal to the difference between the pressure of the steam and that of the atmosphere. Thus the piston R of the forcing pump is being drawn up, and the piston P' is forcing the piston R' down, and thereby driving water into the force pipe S. On the arrival of the piston P at the bottom of the cylinder C, and P' at the top of the cylinder C', the position of the cock is changed as represented in fig. 50. The steam, which has just pressed up the piston P', is allowed to escape into the atmosphere, while the steam, passing from the boiler below the piston P, presses it up, and thus P ascends by the steam pressure, and P' descends by its own weight. By these means the piston R is forced down, driving before it the water in the pump cylinder into the force pipe S, and the piston R' is drawn up to allow the other pump cylinder to be refilled; and so the process is continued.

A valve is placed in the bottom of the force pipes, to prevent the water which has been driven into it from returning. This valve opens upward; and, consequently, the weight of the water pressing upon it only keeps it more effectually closed. On each descent of the piston, the pressure transmitted to the valve acting upward being greater than the weight of the water resting upon it, forces it open, and an increased quantity of water is introduced.

(82.) From the date of the improvement of Watt until

the commencement of the present century, high pressure engines were altogether neglected in these countries. In the year 1802, Messrs. *Trevithick* and *Vivian* constructed the first high-pressure engine which was ever brought into extensive practical use in this kingdom. A section of this machine, made by a vertical plane, is represented in fig. 51.

The boiler A B is a cylinder with flat circular ends. The fireplace is constructed in the following manner:—A tube enters the cylindrical boiler at one end; and, proceeding onward, near the other extremity, is turned and recurved, so as to be carried back parallel to the direction in which it entered. It is thus conducted out of the boiler, at another part of the same end at which it entered. One of the ends of this tube communicates with the chimney E, which is carried upward, as represented in the figure. The other mouth is furnished with a door; and in it is placed the grate, which is formed of horizontal bars, dividing the tube into two parts; the upper part forming the fireplace, and the lower the ash-pit. The fuel is maintained in a state of combustion, on the bars, in that part of the tube represented at C D; and the flame is carried by the draft of the chimney round the curved flue, and issues at E into the chimney. The flame is thus conducted through the water, so as to expose the latter to as much heat as possible.

A section of the cylinder is represented at F, immersed in the boiler, except a few inches of the upper end, where the four-way cock G is placed for regulating the admission of the steam. A tube is represented at H, which leads from this four-way cock into the chimney; so that the waste steam, after working the piston, is carried off through this tube, and passes into the chimney. The upper end of the piston-rod is furnished with a cross-bar, which is placed in a direction at right angles to the length of the boiler, and also to the piston-rod. This bar is guided in its motion by sliding on two iron perpendicular rods fixed to the sides of the boiler, and parallel to each other. To the ends of this cross-bar are joined two connecting rods, the lower ends of which work two cranks fixed on an axis extending across and beneath the boiler, and immediately under the centre of the cylinder. This axis is sustained in bearings formed in the legs which support the boiler, and upon its extremity is fixed the fly wheel as represented at B. A large-toothed wheel is placed on this axis, which, being turned with the cranked axle, communicates motion to other wheels; and, through them, to any machinery which the engine may be applied to move.

As the four-way cock is represented in the figure, the steam passes from the boiler through the curved passage G above the piston, while the steam below the piston is carried off through a tube which does not appear in the figure, by which it is conducted to the tube H, and thence to the chimney. The steam, therefore, which passes above the piston presses it downward; while the pressure upward does not exceed that of the atmosphere. The piston will therefore descend with a force depending on the excess of the pressure of the steam produced in the boiler above the atmospheric pressure. When the piston has arrived at the bottom of the cylinder, the cock is made to assume the position represented in the figure 52. This effect is produced by the motion of the piston-rod. The steam now passes from above the piston, through the tube H, into the chimney, while the steam from the boiler is conducted through another tube below the piston. The pressure above the piston, in this case, does not exceed that of the atmosphere; while the pressure below it will be that of the steam in the boiler. The piston will therefore ascend with the difference of these pressures. On the arrival of the piston at the top of the cylinder, the four-way cock is again turned to the position represented in fig. 51, and the piston again descends; and in the same manner the process is continued. A safety valve is placed on the boiler at V, loaded with a weight W, proportionate to the strength of the steam with which it is proposed to work.

In the engines now described, this valve was frequently loaded at the rate of from 60 to 80 lbs. on the square inch. As the boilers of high-pressure engines were considered more liable to accidents from bursting than those in which steam of a lower pressure was used, greater precautions were taken against such effects. A second safety valve was provided, which was not left in the power of the engine-man. By this means he had a power to diminish the pressure of the steam, but could not increase it beyond the limit determined by the valve, which was removed from his interference. The greatest cause of danger, however, arose from the water in the boiler being consumed by evaporation faster than it was supplied, and therefore falling below the level of the tube containing the furnace. To guard against accidents arising from this circumstance, a hole was bored in the boiler, at a certain depth, below which the water should not be allowed to fall; and in this hole a plug of metal was soldered with lead, or with some other metal, which would fuse at that temperature which would expose the boiler to danger. Thus, in the event of the water being exhausted, so that its level would fall below the plug, the heat of the furnace would immediately melt the solder, and the plug would fall out, affording a vent for the steam, without allowing the boiler to burst. The mercurial steam gauge, already described, was also used as an additional security. When the force of the steam exceeded the length of the column of mercury which the tube would contain, the mercury would be blown out, and the tube would give vent to the steam. The water by which the boiler was replenished was forced into it by a pump worked by the engine. In order to economize the heat, this water was contained in a tube T, which surrounded the pipe H. As the waste steam, after working the piston, passed off through H, it imparted a portion of its heat to the

Pl. XI.

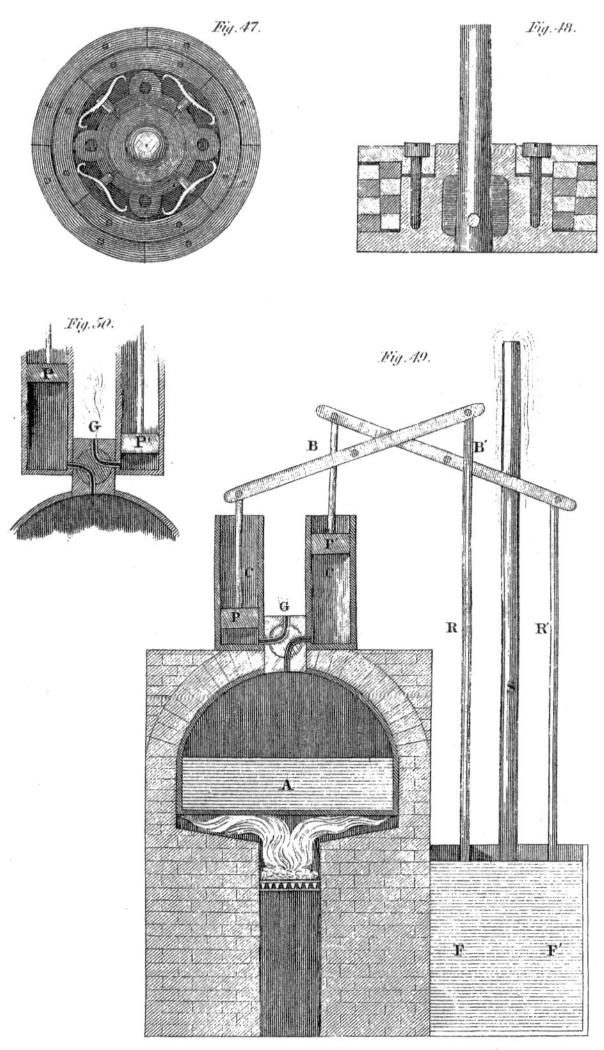


Engr. by P. Maverick

water contained in the tube T, which was thus warmed to a certain temperature before it was forced into the boiler by the pump. Thus a part of the heat, which was originally carried from the boiler in the form of steam, was returned again to the boiler with the water with which it was fed.

It is evident that engines constructed in this manner may be applied to all the purposes to which the condensing engines are applicable.

(*e*) To the plates of the English edition has been added one (plate A) representing a high-pressure engine, as constructed by the West Point Foundry in the state of New York. The principal parts will be readily distinguished from their resemblance to the analogous parts of a condensing engine. The condenser and air-pump of that engine, it will be observed, are suppressed. At $v$ $x$ and $y$ $z$ are forcing pumps, by which a supply of water is injected into the boiler at each motion of the engine. For the four-way cock, used in the English high-pressure engines, a slide valve at $r$ $s$ is substituted, and is found to work to much greater advantage. It is set in motion by an eccentric, in a manner that will be more obvious from an inspection of the plate than from any description.—A. E.

(*f*) A very safe and convenient boiler for a high-pressure engine has been invented in the United States by Mr. Babcock. The boiler consists of small tubes, into which water is flashed by a small forcing pump at every stroke of the engine. The tubes are kept so hot in a furnace as to generate steam of the required temperature, but not hot enough to cause any risk of the decomposition of the water. The strength of the apparatus is such, and the quantity of water exposed to heat at one time so small, as to leave hardly any risk of danger.—A. E.

(83.) Two years after the date of the patent of this engine, its inventor constructed a machine of the same kind for the purpose of moving carriages on railroads; and applied it successfully, in the year 1804, on the railroad at Merthyr

Tydvil, in South Wales. It was in principle the same as that already described. The cylinder, however, was in a horizontal position, the piston-rod working in the direction of the line of road: the extremity of the piston-rod, by means of a connecting rod, worked cranks placed on the axletree, on which were fixed two cogged wheels: these worked in others, by which their motion was communicated finally to cogged wheels fixed on the axle of the hind wheels of the carriage, by which this axle was kept in a state of revolution. The hind wheels being fixed on the axletree, and turning with it, were caused likewise to revolve; and so long as the weight of the carriage did not exceed that which the friction of the road was capable of propelling, the carriage would thus be moved forward. On this axle was placed a fly wheel to continue the rotatory motion at the termination of each stroke. The fore wheels are described as being capable of turning like the fore wheels of a carriage, so as to guide the vehicle. The projectors appear to have contemplated, in the first instance, the use of this carriage on turnpike roads; but that notion seems to have been abandoned, and its use was only adopted on the railroad before mentioned. On the occasion of its first trial, it drew after it as many carriages as contained 10 tons of iron a distance of nine miles; which stage it performed without any fresh supply of water, and travelled at the rate of five miles an hour.

(84.) Capital and skill have of late years been directed with extraordinary energy to the improvement of inland transport; and this important instrument of national wealth and civilization has received a proportionate impulse. Effects are now witnessed, which, had they been narrated a few years since, could only have been admitted into the pages of fiction, or volumes of romance. Who could have credited the possibility of a ponderous engine of iron, loaded with several hundred passengers, in a train of carriages of corresponding magnitude, and a large quantity of water and coal, taking flight from Manchester and arriving at Liverpool, a distance of about thirty miles, in little more than an hour? And yet this is a matter of daily and almost hourly occurrence. Neither is the road, on which this wondrous performance is effected, the most favourable which could be constructed for such machines. It is subject to undulations and acclivities, which reduce the rate of speed much more than similar inequalities affect the velocity on common roads. The rapidity of transport thus attained is not less wonderful than the weights transported. Its capabilities in this respect far transcend the exigencies even of the two greatest commercial marts in Great Britain. Loads varying from 50 to 150 tons are transported at the average rate of fifteen miles an hour; but the engines in this case are loaded below their power; and in one instance we have seen a load—we should rather say a *cargo*—of wagons, conveying merchandise to the amount of 230 tons gross, transported from Liverpool to Manchester at the average rate of twelve miles an hour.

The astonishment with which such performances must be viewed might be qualified, if the art of transport by steam on railways had been matured, and had attained that full state of perfection which such an art is always capable of receiving from long experience, aided by great scientific knowledge, and the unbounded application of capital. But such is not the present case. The art of constructing locomotive engines, so far from having attained a state of maturity, has not even emerged from its infancy. So complete was the ignorance of its powers which prevailed, even among engineers, previous to the opening of the Liverpool railway, that the transport of heavy goods was regarded as the chief object of the undertaking, and its principal source of revenue. The incredible speed of transport, effected even in the very first experiments in 1830, burst upon the public, and on the scientific world, with all the effect of a new and unlooked for phenomenon. On the unfortunate occasion which deprived this country of Mr. Huskisson, the wounded body of that statesman was transported a distance of about fifteen miles in twenty-five minutes, being at the rate of thirty-six miles an hour. The revenue of the road arising from passengers since its opening has, contrary to all that was foreseen, been nearly double that which has been derived from merchandise. So great was the want of experience in the construction of engines, that the company was at first ignorant whether they should adopt large steam engines fixed at different stations on the line, to pull the carriages from station to station, or travelling engines to drag the loads the entire distance. Having decided on the latter, they have, even to the present moment, laboured under the disadvantage of the want of that knowledge which experience alone can give. The engines have been constantly varied in their weight and proportions, in their magnitude and form, as the experience of each successive month has indicated. As defects became manifest they were remedied; improvements suggested were adopted; and each quarter produced engines of such increased power and efficiency, that their predecessors were abandoned, not because they were worn out, but because they had been outstripped in the rapid march of improvement. Add to this, that only one species of travelling engine has been effectively tried; the capabilities of others remain still to be developed; and even that form of engine which has received the advantage of a course of experiments on so grand a scale to carry it toward perfection, is far short of this point, and still has defects, many of which, it is obvious, time and experience will remove. If then travelling steam engines, with all the imperfections of an incipient invention —with the want of experience, the great parent of practical improvements—with the want of the common advantage of the full application of the skill and capital of the country— subjected to but one great experiment, and that experiment limited to one form of engine; if, under such disadvantages, the effects to which we have referred have been produced, what may we not expect from this extraordinary power when the enterprise of the country shall be unfettered, when greater fields of experience are opened, when time, ingenuity, and capital have removed the existing imperfections, and have brought to light new and more powerful principles? This is not mere speculation on possibilities, but refers to what is in a state of actual progression. Railways are in progress between the points of greatest intercourse in the United Kingdom, and travelling steam engines are in preparation for the common turnpike roads; the practicability and utility of that application of the steam engine having not only been established by experiment to the satisfaction of their projectors, but proved before the legislature in a committee of inquiry on the subject.

The important commercial and political effects attending such increased facility and speed in the transport of persons and goods, are too obvious to require any very extended notice here. A part of the price (and in many cases a considerable part) of every article of necessity or luxury consists of the cost of transporting it from the producer to the consumer; and consequently every abatement or saving in this cost must produce a corresponding reduction in the price of every article transported; that is to say, of every thing which is necessary for the subsistence of the poor or for the enjoyment of the rich, of every comfort and of every luxury of life. The benefit of this will extend, not to the consumer only, but to the producer: by lowering the expense of transport of the produce, whether of the soil or of the loom, a less quantity of that produce will be spent in bringing the remainder to market, and consequently a greater surplus will reward the labour of the producer. The benefit of this will be felt even more by the agriculturist than by the manufacturer; because the proportional cost of transport of the produce of the soil is greater than that of manufactures. If 200 quarters of corn be necessary to raise 400, and 100 more be required to bring the 400 to market, then the net surplus will be 100. But if by the use of steam carriages the same quantity can be brought to market with an expenditure of 50 quarters, then the net surplus will be increased from 100 to 150 quarters; and either the profit of the farmer or the rent of the landlord must be increased by the same amount.

But the agriculturist would not merely be benefited by an increased return from the soil already under cultivation. Any reduction in the cost of transporting the produce to market would call into cultivation tracts of inferior fertility, the returns from which would not at present repay the cost of cultivation and transport. Thus land would become productive which is now waste, and an effect would be produced equivalent to adding so much fertile soil to the present extent of the country. It is well known, that land of a given degree of fertility will yield increased produce by the increased application of capital and labour. By a reduction in the cost of transport, a saving will be made which may enable the agriculturists to apply to tracts already under cultivation the capital thus saved, and thereby increase their actual production. Not only, therefore, would such an effect be attended with an increased extent of cultivated land, but also with an increased degree of cultivation in that which is already productive.

It has been said, that in Great Britain there are above a million of horses engaged in various ways in the transport of passengers and goods, and that to support each horse requires as much land as would, upon an average, support eight men. If this quantity of animal power were displaced by steam engines, and the means of transport drawn from the bowels of the earth, instead of being raised upon its surface, then, supposing the above calculation correct, as much land would become available for the support of human beings as would suffice for an additional population of eight millions; or, what amounts to the same, would increase the means of support of the present population by about one-third of the present available means. The land which now supports horses for transport would then support men, or produce corn for food.

The objection that a quantity of land exists in the country capable of supporting horses alone, and that such land would be thrown out of cultivation, scarcely deserves notice here. The existence of any considerable quantity of such land is extremely doubtful. What is the soil which will feed a horse, and not feed oxen or sheep, or produce food for man? But even if it be admitted that there exists in the country a small portion of such land, that portion cannot exceed, nor indeed equal, what would be sufficient for the number of horses which must after all continue to be employed for the purposes of pleasure, and in a variety of cases where steam must necessarily be inapplicable. It is to be remembered, also, that the displacing of horses in one extensive occupation, by diminishing their price, must necessarily increase the demand for them in others.

The reduction in the cost of transport of manufactured articles, by lowering their price in the market, will stimulate their consumption. This observation applies of course not only to home but to foreign markets. In the latter, we already in many branches of manufacture command a monopoly. The reduced price which we shall attain by cheapness and facility of transport will still further extend and increase our advantages. The necessary consequence will be, an increased demand for manufacturing population; and this increased population again reacting on the agricultural interests, will form an increased market for that species of produce. So interwoven and complicated are the fibres which form the texture of the highly civilized and artificial community in which we live, that an effect produced on any one point is instantly transmitted to the most remote and apparently unconnected parts of the system.

The two advantages of increased cheapness and speed, besides extending the amount of existing traffic, call into existence new objects of commercial intercourse. For the same reason that the reduced cost of transport, as we have shown, calls new soils into cultivation, it also calls into existence new markets for manufactured and agricultural produce. The great speed of transit which has been proved to be practicable must open a commerce between distant points in various articles, the nature of which does not permit them to be preserved so as to be fit for use beyond a certain time. Such are, for example, many species of vegetable and animal food, which at present are confined to markets at a very limited distance from the grower or feeder. The truth of this observation is manifested by the effects which have followed the intercourse by steam on the Irish Channel. The western towns of England have become markets for a prodigious quantity of Irish produce, which it had been previously impossible to export. If animal food be transported alive from the grower to the consumer, the distance of the market is limited by the power of the animal to travel, and the cost of its support on the road. It is only particular species of cattle which bear to be carried to market on common roads and by horse carriages. But the peculiar nature of a railway, the magnitude and weight of the loads which may be transported on it, and the prodigious speed which may be attained, render the transport of cattle, of every species, to almost any distance, both easy and cheap. In process of time, when the railway system becomes extended, the metropolis and populous towns will therefore become markets, not as at present to districts within limited distances of them, but to the whole country.

The moral and political consequences of so great a change in the powers of transition of persons and intelligence from place to place are not easily calculated. The concentration of mind and exertion which a great metropolis always exhibits, will be extended in a considerable degree to the whole realm. The same effect will be produced as if all distances were lessened in the proportion in which the speed and cheapness of transit are increased. Towns, at present removed some stages from the metropolis, will become its suburbs; others, now a day's journey, will be removed to its immediate vicinity; business will be carried on with as much ease between them and the metropolis, as it is now between distant points of the metropolis itself. Let those who discard speculations like these as wild and improbable, recur to the state of public opinion, at no very remote period, on the subject of steam navigation. Within the memory of persons who have not yet passed the meridian of life, the possibility of traversing by the steam engine the channels and seas that surround and intersect these islands was regarded as the dream of enthusiasts. Nautical men and men of science rejected such speculations with equal incredulity, and with little less than scorn for the understanding of those who could for a moment entertain them. Yet we have witnessed steam engines traversing not these channels and seas alone, but sweeping the face of the waters round every coast in Europe. The seas which interpose between our Asiatic dominions and Egypt, and those which separate our own shores from our West Indian possessions, have offered an equally ineffectual barrier to its powers. Nor have the terrors of the Pacific prevented the "Enterprise" from doubling the Cape, and reaching the shores of India. If steam be not used as the only means of connecting the most distant points of our planet, it is not because it is inadequate to the accomplishment of that end, but because the supply of the material from which at the present moment it derives its powers, is restricted by local and accidental circumstances.*

We propose in the present chapter to lay before our readers some account of the means whereby the effects above referred to have been produced; of the manner and degree in which the public have availed themselves of these means;

* Some of the preceding observations on inland transport, as well as other parts of the present chapter, appeared in articles written by me in the Edinburgh Review for October, 1832, and October, 1834.

and of the improvements of which they seem to us to be susceptible.

(85.) It is a singular fact, that in the history of this invention, considerable time and great ingenuity were vainly expended in attempting to overcome a difficulty which in the end turned out to be purely imaginary. To comprehend distinctly the manner in which a wheel carriage is propelled by steam, suppose that a pin or handle is attached to the spoke of the wheel at some distance from its centre, and that a force is applied to this pin in such a manner as to make the wheel revolve. If the face of the wheel and the surface of the road were absolutely smooth and free from friction, so that the face of the wheel would slide without resistance upon the road, then the effect of the force thus applied would be merely to cause the wheel to turn round, the carriage being stationary, the surface of the wheel would slip or slide upon the road as the wheel is made to revolve. But if, on the other hand, the pressure of the face of the wheel upon the road is such as to produce between them such a degree of adhesion as will render it impossible for the wheel to slide or slip upon the road by the force which is applied to it, the consequence will be, that the wheel can only turn round in obedience to the force which moves it by causing the carriage to advance, so that the wheel will roll upon the road, and the carriage will be moved forward, through a distance equal to the circumference of the wheel, each time it performs a complete revolution.

It is obvious that both of these effects may be partially produced; the adhesion of the wheel to the road may be insufficient to prevent slipping altogether, and yet it may be sufficient to prevent the wheel from slipping as fast as it revolves. Under such circumstances the carriage would advance, and the wheel would slip. The progressive motion of the carriage during one complete revolution of the wheel would be equal to the difference between the complete circumference of the wheel and the portion through which in one revolution it has slipped.

When the construction of travelling steam engines first engaged the attention of engineers, and for a considerable period afterward, a notion was impressed upon their minds that the adhesion between the face of the wheel and the surface of the road must necessarily be of very small amount, and that in every practical case the wheels thus driven would either slip altogether, and produce no advance of the carriage, or that a considerable portion of the impelling power would be lost by the partial slipping or sliding of the wheels. It is singular that it should never have occurred to the many ingenious persons, who for several years were engaged in such experiments and speculations, to ascertain by experiment the actual amount of adhesion in any particular case between the wheels and the road. Had they done so, we should probably now have found locomotive engines in a more advanced state than that to which they have attained.

To remedy this imaginary difficulty, Messrs. *Trevithick* and *Vivian* proposed to make the external rims of the wheels rough and uneven, by surrounding them with projecting heads of nails or bolts, or by cutting transverse grooves on them. They proposed, in cases where considerable elevations were to be ascended, to cause claws or nails to project from the surface during the ascent, so as to take hold of the road.

In seven years after the construction of the first locomotive engine by these engineers, another locomotive engine was constructed by Mr. *Blinkensop,* of Middleton Colliery, near Leeds. He obtained a patent, in 1811, for the application of a rack-rail. The railroad thus, instead of being composed of smooth bars of iron, presented a line of projecting teeth, like those of a cog-wheel, which stretched along the entire distance to be travelled. The wheels on which the engine rolled were furnished with corresponding teeth, which worked in the teeth of the railroad; and, in this way, produced a progressive motion in the carriage.

The next contrivance for overcoming this fictitious difficulty was that of Messrs. *Chapman*, who, in the year 1812, obtained a patent for working a locomotive engine by a chain extending along the middle of the line of railroad, from the one end to the other. This chain was passed once round a grooved wheel under the centre of the carriage; so that, when this grooved wheel was turned by the engine, the chain being incapable of slipping upon it, the carriage was consequently advanced on the road. In order to prevent the strain from acting on the whole length of the chain, its links were made to fall upon upright forks placed at certain intervals, which between those intervals sustained the tension of the chain produced by the engine. Friction rollers were used to press the chain into the groove of the wheel, so as to prevent it from slipping. This contrivance was soon abandoned, for the very obvious reason that a prodigious loss of force was incurred by the friction of the chain.

The following year, 1813, produced a contrivance of singular ingenuity, for overcoming the supposed difficulty arising from the want of adhesion between the wheels and the road. This was no other than a pair of mechanical legs and feet, which were made to walk and propel in a manner somewhat resembling the feet of an animal.

A sketch of these propellers is given in fig. 53. A is the

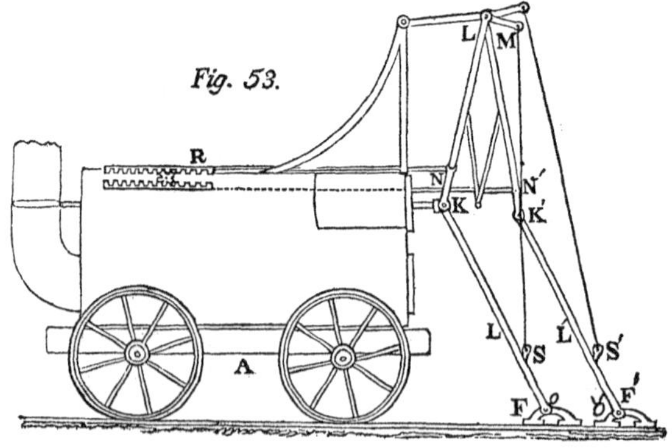

*Fig. 53.*

carriage moving on the railroad, L and L′ are the legs, F and F′ the feet. The foot F has a joint at O, which corresponds to the ankle; another joint is placed at K, which corresponds to the knee; and a third is placed at L, which corresponds to the hip. Similar joints are placed at the corresponding letters in the other leg. The knee joint K is attached to the end of the piston of the cylinder. When the piston, which is horizontal, is pressed outward, the leg L presses the foot F against the ground, and the resistance forces the carriage A onwards. As the carriage proceeds, the angle K at the knee becomes larger, so that the leg and thigh take a straighter position; and this continues until the piston has reached the end of its stroke. At the hip L there is a short lever L M, the extremity of which is connected by a cord or chain with a point S, placed near the shin of the leg. When the piston is pressed into the cylinder, the knee K is drawn toward the engine, and the cord M S is made to lift the foot F from the ground; to which it does not return until the piston has arrived at the extremity of the cylinder. On the piston being again driven out of the cylinder, the foot F, being placed on the road, is pressed backward by the force of the piston-rod at K; but the friction of the ground preventing its backward motion, the reaction causes the engine to advance: and in the same manner this process is continued.

Attached to the thigh at N, above the knee, by a joint, is a horizontal rod N R, which works a rack R. This rack has beneath it a cog-wheel. This cog-wheel acts in another rack below it. By these means, when the knee K is driven *from* the engine, the rack R is moved *backward;* but the cog-wheel, acting on the other rack beneath it, will move the latter *in the contrary direction.* The rack R being then moved *in the same direction with the knee* K, it follows that the other rack will always be moved in a *contrary direction.* The lower rack is connected by another horizontal rod with the thigh of the leg L F′, immediately above the knee at N′. When the piston is forced *inward,* the knee K' will thus be forced *backward;* and when the piston is forced *outward,* the knee K' will be drawn *forward.* It therefore follows that the two knees, K and K', are pressed *alternately backward* and *forward.* The foot F', when the knee K' is drawn forward, is lifted by the means already described for the foot F.

It will be apparent, from this description, that the piece of mechanism here exhibited is a contrivance derived from the motion of the legs of an animal, and resembling in all respects the fore legs of a horse. It is however to be regarded rather as a specimen of great ingenuity than as a contrivance of practical utility.

(86.) It was about this period that the important fact was first ascertained that the adhesion or friction of the wheels with the rails on which they moved was amply sufficient to propel the engine, even when dragging after it a load of great weight; and that in such case, the progressive motion would be effected without any slipping of the wheels. The consequence of this fact rendered totally useless all the contrivances for giving wheels a purchase on the road, such as racks, chains, feet, &c. The experiment by which this was determined appears to have been first tried on the Wylam railroad; where it was proved, that, when the road was level, and the rails clean, the adhesion of the wheels was sufficient, in all kind of weather, to propel considerable loads. By manual labour it was first ascertained how much weight the wheels of a common carriage would overcome without slipping round on the rail; and having found the proportion which that bore to the weight, they then ascertained that the weight of the engine would produce sufficient adhesion to drag after it, on the railroad, the requisite number of wagons.*

In 1814, an engine was constructed at Killingworth, by Mr. *Stephenson,* having two cylinders with a cylindrical boiler, and working two pair of wheels, by cranks placed at

* Wood on Railroads, 2d edit.

right angles; so that when the one was in full operation, the other was at its dead points. By these means the propelling power was always in action. The cranks were maintained in this position by an endless chain, which passed round two cogged wheels placed under the engine, and which were fixed on the same axles on which the wheels were placed. The wheels in this case were fixed on the axles, and turned with them.

This engine is represented in fig. 54, the sides being open, to render the interior mechanism visible. A B is the cylindrical boiler; C C are the working cylinders; D E are the

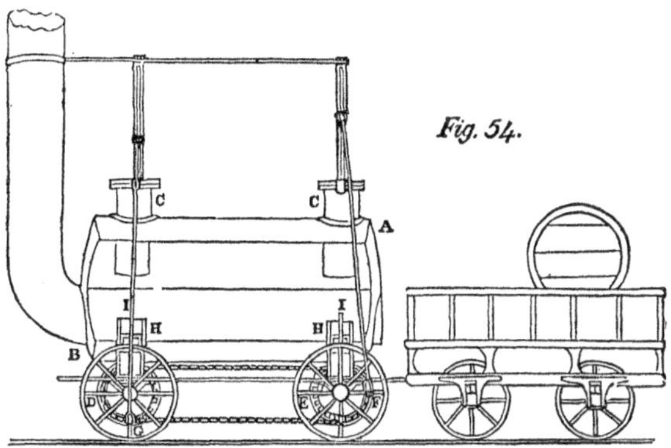


cogged wheels fixed on the axle of the wheels of the engine, and surrounded by the endless chain. These wheels, being equal in magnitude, perform their revolutions in the same time; so that, when the crank F descends to the lowest point, the crank G rises from the lowest point to the horizontal position D; and, again, when the crank F rises from the lowest point to the horizontal position E, the other crank rises to the highest point; and so on. A very beautiful contrivance was adopted in this engine, by which it was suspended on springs of steam. Small cylinders, represented at H, are screwed by flanges to one side of the boiler, and project within it a few inches: they have free communica tion at the top with the water or steam of the boiler. Solid pistons are represented at I, which move steam-tight in these cylinders; the cylinders are open at the bottom, and the piston-rods are screwed on the carriage of the engine, over the axle of each pair of wheels, the pistons being presented upward. As the engine is represented in the figure, it is supported on four pistons, two at each side. The pistons are pressed upon by the water or steam which occupies the upper chamber of the cylinder; and the latter being elastic in a high degree, the engine has all the advantage of spring suspension. The defect of this method of supporting the engine is, that when the steam loses that amount of elasticity necessary for the support of the machine, the pistons are orced into the cylinders, and the bottoms of the cylinders bear upon them. All spring suspension is then lost. This mode of suspension has consequently since been laid aside.

In an engine subsequently constructed by Mr. *Stephenson*, for the Killingworth railroad, the mode adopted of connecting the wheels by an endless chain and cog-wheels was abandoned; and the same effect was produced by connecting the two cranks by a straight rod. All such contrivances, however, have this great defect, that, if the fore and hind wheels be not constructed with dimensions accurately equal, there must necessarily be a slipping or dragging on the road. The nature of the machinery requires that each wheel should perform its revolution exactly in the same time; and consequently, in doing so, must pass over exactly equal lengths of the road. If, therefore, the circumference of the wheels be not accurately equal, that wheel which has the lesser circumference must be dragged along so much of the road as that by which it falls short of the circumference of the greater wheel; or, on the other hand, the greater must be dragged in the opposite direction, to compensate for the same difference. As no mechanism can accomplish a perfect equality in four, much less in six, wheels, it may be assumed that a great portion of that dragging effect is a necessary consequence of the principle of this machine; and even were the wheels, in the first instance, accurately constructed, it is not possible that their wear could be so exactly uniform as to continue equal.

(87.) The next stimulus which the progress of this invention received, proceeded from the great national work undertaken at Liverpool, by which that town and the extensive commercial mart of Manchester were connected by a double line of railway. When this project was undertaken, it was not decided what moving power it might be most expedient to adopt as a means of transport on the proposed road: the choice lay between horse power, fixed steam engines, and locomotive engines; but the first, for many obvious reasons, was at once rejected in favour of one or other of the last two.

The steam engine may be applied, by two distinct methods, to move wagons either on a turnpike road or on a railway. By the one method the steam engine is fixed, and draws the carriage or train of carriages toward it by a chain extending the whole length of road on which the engine works. By this method the line of road over which the transport is conducted is divided into a number of short intervals, at the extremity of each of which an engine is placed. The wagons or carriages, when drawn by any engine to its own station, are detached, and connected with the extremity of the chain worked by the next stationary engine; and thus the journey is performed, from station to station, by separate engines. By the other method, the same engine draws the load the whole journey, travelling with it.

The Directors of the Liverpool and Manchester railroad, when that work was advanced toward its completion, employed, in the spring of the year 1829, Messrs. *Stephenson* and *Lock*, and Messrs. *Walker* and *Rastrick*, experienced engineers, to visit the different railways where practical information respecting the comparative effects of stationary and locomotive engines was likely to be obtained; and from these gentlemen they received reports on the relative merits, according to their judgment, of the two methods. The particulars of their calculations are given at large in the valuable work of Mr. *Nicholas Wood* on railways: to which we refer the reader, not only on this, but on many other subjects connected with the locomotive steam engine, into which it would be foreign to our subject to enter. The result of the comparison of the two systems was, that the capital necessary to be advanced to establish a line of stationary engines was considerably greater than that which was necessary to establish an equivalent power in locomotive engines; that the annual expense by the stationary engines was likewise greater; and that, consequently, the expense of transport by the latter was greater, in a like proportion. The subjoined table exhibits the results numerically :—

| | Capital. | Annual Expense. | Expense of taking a Ton of Goods One Mile. |
|---|---|---|---|
| | £ s. d. | £ s. d. | |
| Locomotive Engines..... | 58,000 0 0 | 25,517 8 2 | 0·164 of a penny. |
| Stationary Engines...... | 121,496 7 0 | 42,031 16 5 | 0·269 |
| Locomotive System—less | 63,496 7 0 | 16,514 8 3 | 0·105 |

On the score of economy, therefore, the system of locomotive engines was entitled to a preference; but there were other considerations which conspired with this to decide the choice of the Directors in its favour. An accident occurring in any part of a road worked by stationary engines must necessarily produce a total suspension of work along the entire line. The most vigilant and active attention on the part of every workman, however employed, in every part of the line, would therefore be necessary; but, independently of this, accidents arising from the fracture or derangement of any of the chains, or from the suspension of the working of any of the fixed engines, would be equally injurious, and would effectually stop the intercourse along the line. On the other hand, in locomotive engines an accident could only affect the particular train of carriages drawn by the engine to which the accident might occur; and even then the difficulty could be remedied by having a supply of spare engines at convenient stations along the line. It is true that the *probability* of accident is, perhaps, less in the stationary than in the locomotive system; but the *injurious consequences*, when accident *does* happen, are prodigiously greater in the former. "The one system," says Mr. *Walker*, "is like a chain extending from Liverpool to Manchester, the failure of a single link of which would destroy the whole; while the other is like a number of short and unconnected chains," the destruction of any one of which does not interfere with the effect of the others, and the loss of which may be supplied with facility.

The decision of the directors was, therefore, in favour of locomotive engines; and their next measure was to devise some means by which the inventive genius of the country might be stimulated to supply them with the best possible form of engines for this purpose. With this view, it was proposed and carried into effect to offer a prize for the best locomotive engine which might be produced under certain proposed conditions, and to appoint a time for a public trial of the claims of the candidates. A premium of 500*l.* was accordingly offered for the best locomotive engine to run on the Liverpool and Manchester railway; under the condition that it should produce no smoke; that the pressure of the steam should be limited to 50lbs. on the inch; and that it should draw at least three times its own weight, at the rate of not less than ten miles an hour; that the engine should be supported on springs, and should not exceed fifteen feet in height. Precautions were also proposed against the consequences of the boiler bursting; and other matters, not necessary to mention more particularly here. This proposal was announced in the spring of 1829, and the time of trial was appointed in the following October. The engines which underwent the trial were, the Rocket, constructed by Mr. *Stevenson;* the Sanspareil, by *Hackworth;* and the Novelty, by Messrs. *Braithwait* and *Ericson.* Of these, the Rocket obtained the premium. A line of railway was selected for the trial, on a level piece of road about two miles in length, near a place called Rainhill, between Liverpool and Manchester; the distance between the two stations was a mile and a half, and the engine had to travel this distance backward and forward ten times, which made altogether a journey of 30 miles. The Rocket performed this journey twice: the first time in 2 hours 14 minutes and 8 seconds; and the second time, in 2 hours 6 minutes and 49 seconds. Its speed at different parts of the journey varied: its greatest rate of motion was rather above 29 miles an hour; and its least, about 11½ miles an hour. The average rate of the one journey was $13\frac{4}{10}$ miles an hour; and of the other, $14\frac{2}{20}$ miles. This was the only engine which performed the complete journey proposed, the others having been stopped from accidents which occurred to them in the experiment. The Sanspareil performed the distance between the stations eight times, travelling 22½ miles in 1 hour 37 minutes and 16 seconds. The greatest velocity to which this engine attained was something less than 23 miles per hour. The Novelty had only passed twice between the stations when the joints of the boiler gave way, and put an end to the experiment.

(88.) The great object to be attained in the construction of these engines was, to combine with sufficient lightness the greatest possible heating power. The fire necessarily acts on the water in two ways: first, by its radiant heat; and second, by the current of heated air which is carried by the draft through the fire, and finally passes into the chimney. To accomplish this object, therefore, it is necessary to expose to both these sources of heat the greatest possible quantity of surface in contact with the water. These ends were attained by the following admirable arrangement in the Rocket:—

This engine is represented in fig. 55. It is supported on four wheels; the principal part of the weight being thrown

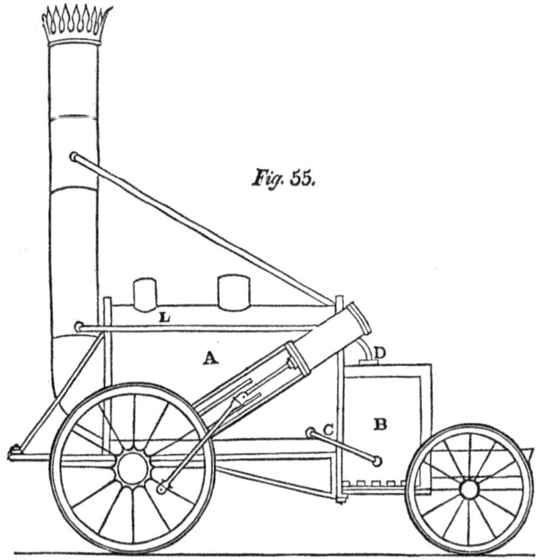


*Fig.* 55.

on one pair, which are worked by the engine. The boiler consists of a cylinder six feet in length, with flat ends; the chimney issues from one end, and to the other end is attached a square box, B, the bottom of which is furnished with the grate on which the fuel is placed. This box is composed of two casings of iron, one contained within the other, having between them a space about 3 inches in breadth; the magnitude of the box being 3 feet in length, 2 feet in width, and 3 feet in depth. The casing which surrounds the box communicates with the lower part of the boiler by a pipe marked C; and the same casing at the top of the box communicates with the upper part of the boiler by another pipe marked D. When water is admitted into the boiler, therefore, it flows freely through the pipe C, into the casing which surrounds the furnace or fire box, and fills this casing to the same level as that which it has in the boiler. When the engine is at work, the boiler is kept about half filled with water; and, consequently, the casing surrounding the furnace is completely filled. The steam which is generated in the water contained in the casing finds its exit through the pipe D, and escapes into the upper part of the boiler. A section of the engine, taken at right angles to its length is represented at fig. 56. Through the lower part of the boiler pass a number of copper tubes of small size, which communicate at one end with the fire box, and at the other with the chimney, and form a passage for the heated air from the furnace to the chimney. The ignited fuel spread on the grate at the bottom of the fire box disperses its heat by radiation, and acts in this manner on the whole surface of the casing surrounding the fire box; and thus raises the temperature of the thin shell of water contained in that casing. The chief part of the water in the casing, being lower in its position than the water in the boiler, acquires a tendency to ascend when heated, and passes into the boiler; so that a constant circulation of the heated water is maintained, and the water in the boiler must necessarily be kept at nearly the same temperature as the water in the casing. The air which passes through the burning fuel, and which fills the fire box, is carried by the draft through the tubes which extend through the lower part of the boiler; and as these tubes are surrounded on every side with the water contained in the boiler, this air transmits its heat through these tubes to the water. It finally issues into the chimney, and rises by the draft. The power of this furnace must necessarily depend on the power of draft in the chimney; and to increase this, and at the same time to dispose of the waste steam after it has worked the piston, this

Fig. 56.

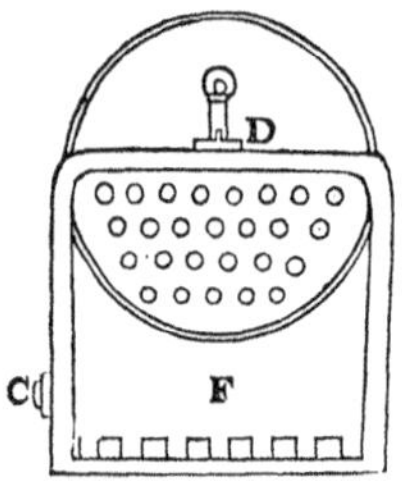

steam is carried off by a pipe L, which passes from the cylinder to the chimney, and escapes there in a jet which is turned upward. By the velocity with which it issues from this jet, and by its great comparative levity, it produces a strong current upward in the chimney, and thus gives force to the draft of the furnace. In fig. 56, the grate bars are represented at the bottom of the fire box at F. There are two cylinders, one of which works each wheel; one only appearing in the drawing, (fig. 55,) the other being concealed by the engine. The spokes which these cylinders work are placed at right angles on the wheels; the wheels being fixed on a common axle, with which they turn.

In this engine, the surface of water surrounding the fire box, exposed to the action of radiant heat, amounted to 20 square feet, which received heat from the surface of 6 square feet of burning fuel on the bars. The surface exposed to the action of the heated air amounted to 118 square feet. The engine drew after it another carriage, containing fuel and water; the fuel used was coke, for the purpose of avoiding the production of smoke.

(89.) The Sanspareil of Mr. *Hackworth* is represented in fig. 57; the horizontal section being exhibited in fig. 58.

The draft of the furnace is produced in the same manner as in the Rocket, by ejecting the waste steam coming from the cylinder into the chimney; the boiler, however, differs considerably from that of the Rocket. A recurved tube passes through the boiler, somewhat similar to that already described in the early engine of Messrs. *Trevithick* and *Vivian*. In the horizontal section, (fig. 58,) D expresses the opening of the furnace at the end of the boiler, beside the chimney. The grate bars appear at A, supporting the burning fuel; and a curved tube passing through the boiler, and terminating in the chimney, is expressed at B, the direction of the draft being indicated by the arrow; C is a section of the chimney. The cylinders are placed, as in the Rocket, on each side of the boiler; each working a separate wheel, but acting on spokes placed at right angles to each other. The tube in which the grate and flue are placed diminishes in diameter as it approaches the chimney. At the mouth where the grate was placed, its diameter was two feet; and it was gradually reduced, so that, at the chimney, its diameter was only fifteen inches. The grate bars extended 5 feet into the tube. The surface of water exposed to the radiant heat

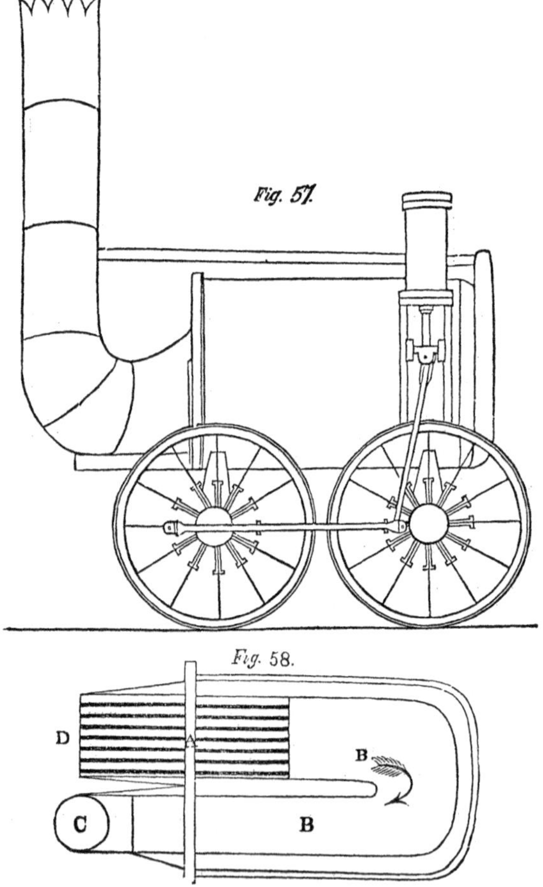

of the fire was 16 square feet; and that exposed to the action of the heated air and flame was about 75 square feet. The magnitude of the grate or sheet of burning fuel which radiated heat, was 10 square feet.

(90.) The Novelty, of Messrs. *Braithwait* and *Ericson*, is represented in fig. 59; and a section of the generator and boiler is exhibited in fig. 60: the corresponding parts in the two figures are marked by the same letters.

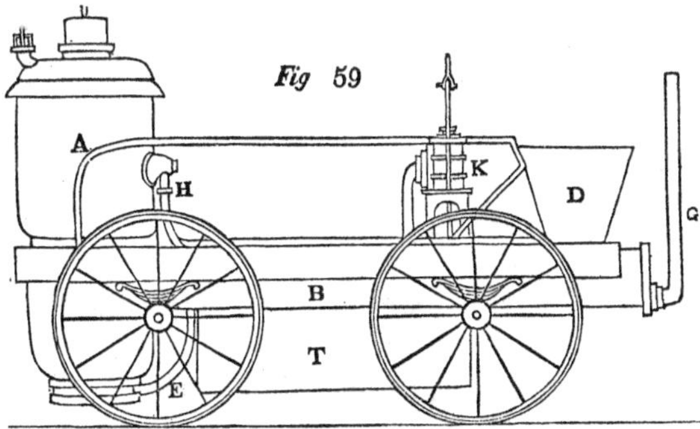


A is the generator or receiver, containing the steam which works the engine; this communicates with a lower generator B, which extends in a horizontal direction the entire length of the carriage. Within the generator A is contained the furnace F, which communicates in a tube C; carried up through the generator, and terminated at the top by sliding shutters, which exclude the air, and which are only opened to supply fuel to the grate F. Below the grate the furnace is not open, as usual, to the atmosphere, but communicates by a tube E, with a bellows D; which is worked by the engine, and which forces a constant stream of air, by the tube E, through the fuel on F, so as to keep that fuel in vivid combustion. The heated air, contained in the furnace F, is driven on, by the same force, through a small curved tube marked *e*, which circulates like a worm, (as represented in fig. 60,) through the horizontal generator or receiver; and, tapering gradually, until reduced to very small dimensions, it finally issues into the chimney G. The air, in passing along this tube, imparts its heat to the water by which the tube is surrounded, and is brought to a considerably reduced

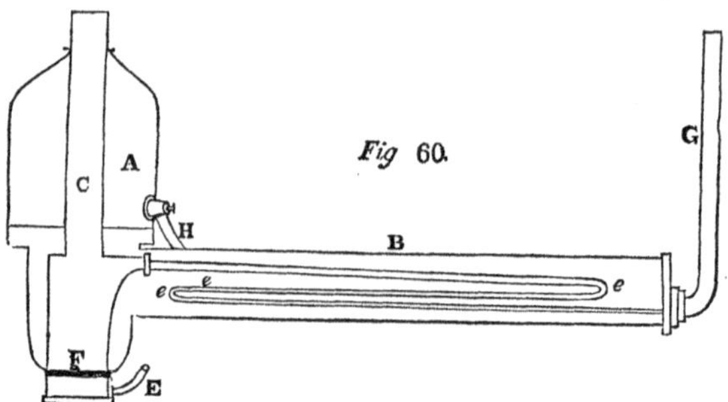


temperature when discharged into the chimney. The cylinder, which is represented at K, works one pair of wheels, by means of a bell crank; the other pair, when necessary, being connected with them.

In this engine, the magnitude of the surface of burning fuel on the grate bars is less than 2 square feet; the surface exposed to radiant heat is 9½ square feet; and the surface of water exposed to heated air is about 33 square feet.

The superiority of the Rocket may be attributed chiefly to the greater quantity of surface of the water which is exposed to the action of the fire. With a less extent of grate bars than the Sanspareil, in the proportion of 3 to 5, it exposes a greater surface of water to radiant heat, in the proportion of 4 to 3; and a greater surface of water to heated air, in the proportion of more than 3 to 2. It was found that the Rocket, compared with the Sanspareil, consumed fuel, in the evaporation of a given quantity of water, in the proportion of 11 to 28. The suggestion of using the tubes to conduct through the water the heated air to the chimney is due to Mr. *Booth*, treasurer of the Liverpool and Manchester Railway Company; and, certainly, nothing has been more conducive to the efficiency of the engines since used than this improvement. It is much to be regretted that the ingenious gentleman who suggested this has reaped none of the advantages to which a patentee would be legally entitled.*

(91.) The great object to be effected in the boilers of these engines is, to keep a small quantity of water at an excessive temperature, by means of a small quantity of fuel kept in the most active state of combustion. To accomplish this, it is necessary, first, so to shape the boiler, furnace, and flues, that the water shall be in contact with as extensive a surface as possible, every part of which is acted on either immediately, by the heat radiating from the fire, or mediately, by the air which has passed through the fire, and which finally rushes into the chimney : and, secondly, that such a forcible draught should be maintained in the furnace, that a quantity of heat shall be extricated from the fuel, by combustion, sufficient to maintain the water at the necessary temperature, and to produce the steam with sufficient rapidity. To accomplish these objects, therefore, the chamber containing the grate should be completely surrounded by water, and should be below the level of the water in the boiler. The magnitude of the surface exposed to radiation should be as great as is consistent with the whole magnitude of the machine. The comparative advantage which the Rocket possessed in these respects over the other engines will be evident on inspection. In the next place, it is necessary that the heat, which is absorbed by the air passing through the fuel, and keeping it in a state of combustion, should be transferred to the water before the air escapes into the chimney. Air being a bad conductor of heat, to accomplish this it is necessary that the air in the flues should be exposed to as great an extent of surface in contact with the water as possible. No contrivance can be less adapted for the attainment of this end than one or two large tubes traversing the boiler, as in the earliest locomotive engines : the body of air which

* Mr. *Booth* received a part of the premium of 500*l.*, but has not participated in any degree in the profits of the manufacture of the engines.

passes through the centre of these tubes had no contact with their surface, and, consequently, passed into the chimney at nearly the same temperature as that which it had when it quitted the fire. The only portion of air which imparted its heat to the water was that portion which passed next to the surface of the tube.

Several methods suggest themselves to increase the surface of water in contact with a given quantity of air passing through it. This would be accomplished by causing the air to pass between plates placed near each other, so as to divide the current into thin strata, having between them strata of water, or it might be made to pass between tubes differing slightly in diameter, the water passing through an inner tube, and being also in contact with the external surface of the outer tube. Such a method would be similar in principle to the steam jacket used in *Watt's* steam engines, or to the condenser of *Cartwright's* engine already described. But, considering the facility of constructing small tubes, and of placing them in the boiler, that method, perhaps, is, on the whole, the best in practice, ; although the shape of a tube, geometrically considered, is most unfavourable for the exposure of a fluid contained in it to its surface. The air which passes from the fire-chamber, being subdivided as it passes through the boiler by a great number of very small tubes, may be made to impart all its excess of heat to the water before it issues into the chimney. This is all which the most refined contrivance can effect. The Rocket engine was traversed by 25 tubes, each 3 inches in diameter; and the principle has since been carried to a much greater extent.

The abstraction of a great quantity of heat from the air before it reaches the chimney is attended with one consequence, which, at first view, would present a difficulty apparently insurmountable ; the chimney would, in fact, lose its power of draught. This difficulty, however, was removed by using the waste steam, which had passed from the cylinder after working the engine, for the purpose of producing a draught. This steam was urged through a jet presented upward in the chimney, and driven out with such force in that direction as to create a sufficient draught to work the furnace.

It will be observed that the principle of draught in the Novelty is totally distinct from this: in that engine the draught is produced by a bellows worked by the engine. The question, as far as relates to these two methods, is, whether more power is lost in supplying the steam through the jet, as in the Rocket, or in working the bellows, as in the Novelty. The force requisite to impel the steam through the jet must be exerted by the returning stroke of the piston, and, consequently, must rob the working effect to an equivalent amount. On the other hand, the power requisite to work the bellows in the Novelty must be subducted from the available power of the engine. The former method is found to be the more effectual and economical.

The importance of these details will be understood, when it is considered that the only limit to the attainment of speed by locomotive engines is the power to produce in a given time a certain quantity of steam. Each stroke of the piston causes one revolution of the wheels, and consumes two cylinders full of steam: consequently, a cylinder of steam corresponds to a certain number of feet of road travelled over: hence it is that the production of a rapid and abundant supply of heat, and the imparting of that heat quickly and effectually to the water, is the key to the solution of the problem to construct an engine capable of rapid motion.

The method of subdividing the flue into tubes was carried much further by Mr. *Stephenson* after the construction of the Rocket; and, indeed, the principle was so very obvious, that it is only surprising that, in the first instance, tubes of smaller diameter than 3 inches were not used. In engines since constructed, the number of tubes vary from 90 to 120, the diameter being reduced to 2 inches or less, and in some instances tubes have been introduced, even to the number of 150, of 1½ inch diameter. In the Meteor, 20 square feet are exposed to radiation, and 139 to the contact of heated air; in the Arrow, 20 square feet to radiation, and 145 to the contact of heated air. The superior economy of fuel gained by this means will be apparent by inspecting the following table, which exhibits the consumption of fuel which was requisite to convey a ton weight a mile in each of four engines, expressing also the rate of the motion :—

| Engines. | Average rate of speed in miles per hour. | Consumption of coke in pounds per ton per mile. |
|---|---|---|
| No. 1. Rocket....... | 14 | 2·41 |
| 2. Sanspareil.... | 15 | 2·47 |
| 3. Phœnix...... | 12 | 1·42 |
| 4. Arrow....... | 12 | 1·25 |

(92.) Since the period at which the railway was opened for the actual purposes of transport, the locomotive engines have been in a state of progressive improvement. Scarcely a month has passed without suggesting some change in the details, by which fuel might be economized, the production of steam rendered more rapid, the wear of the engine rendered slower, the proportionate strength of the different parts improved, or some other desirable end obtained. The consequence of this has been, that the particular engines to which we have alluded, and others of the same class, without having, as it were, lived their natural life, or without having been worn out by work, have been laid aside to give place to others of improved powers. By the exposure of the cylinders to the atmosphere in the Rocket, and engines of a similar form, a great waste of heat was incurred, and it was accordingly determined to remove them from the exterior of the boiler, and to place them within a casing immediately under the chimney : this chamber was necessarily kept warm by its proximity to the end of the boiler, but more by the current of heated air which constantly rushed into it from the tubes. This change, also, rendered necessary another, which improved the working of the engine. In the earlier engines the motion of the piston was communicated to the wheel by a connecting rod attached to one of the spokes on the exterior of the wheel, as represented in fig. 55. By the change to which we have just alluded, the cylinders being placed between the wheels under the chimney, this mode of working became inapplicable, and it was considered better to connect the piston-rods with two cranks placed at right angles on the axles of the great wheels. By this means, it was found that the working of the machine was more even, and productive of less strain than in the former arrangement. On the other hand, a serious disadvantage was incurred by the adoption of a cranked axle. The weakness necessarily arising from such a form of axle could only be counterbalanced by great thickness and weight of metal; and even this precaution does not prevent the occasional fracture of such axles at the angles of the cranks. The advantages, however, of this plan, on the whole, are considered to predominate.

In the most improved engines in present use two safety valves are provided, of which only one is in the power of the engine-man. The tubes being smaller and more numerous than in the earlier engines, the heat is more completely extracted from the air before it enters the chimney. A powerful draft is rendered still more necessary by the smallness of the tubes: this is effected by forcing the steam which has worked the pistons through a contracted orifice, presented upward in the chimney, by the regulation of which any degree of draft may be obtained.

One of the most improved engines at present in use is represented in fig. 61.

A represents the cylindrical boiler, the lower half of which is traversed by tubes, as described in the Rocket. They are usually from 80 to 100 in number, and about 1½ inch in diameter; the boiler is about 7 feet in length; the fire-cham

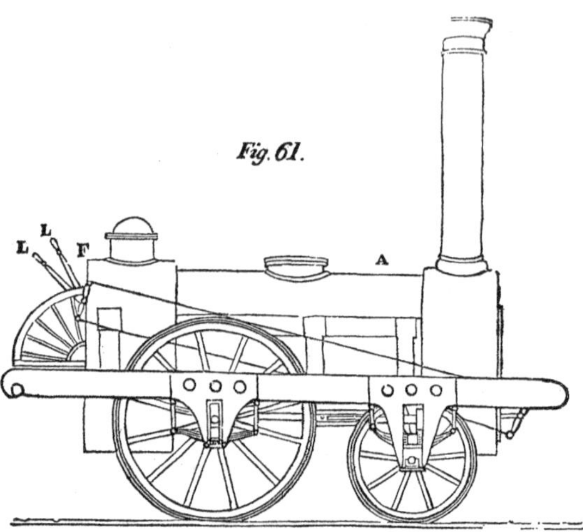

Fig. 61.

ber is attached to one end of it, at F, as in the Rocket, and similar in construction; the cylinders are inserted in a chamber at the other end, immediately under the chimney. The piston-rods are supported in the horizontal position by guides; and connecting rods extend from them, under the engine, to the two cranks placed on the axle of the large wheels. The effects of any inequality in the road are counteracted by springs, on which the engine rests; the springs being below the axle of the great wheels, and above that of the less. The steam is supplied to the cylinders, and withdrawn, by means of the common sliding valves, which are worked by an eccentric wheel placed on the axle of the large wheels of the carriage. The motion is communicated from this eccentric wheel to the valve by sliding rods. The stand is placed for the attendant at the end of the engine, next the fireplace F; and two levers L, project from the end, which communicate with the valves by means of rods, by which the engine is governed, so as to stop or reverse the motion.

The wheels of these engines have been commonly constructed of wood, with strong iron tires, furnished with flanges adapted to the rails. But Mr. *Stephenson* has recently substituted, in some instances, wheels of iron with hollow spokes. The engine draws after it a tender carriage containing the fuel and water; and, when carrying a light load, is capable of performing the whole journey from Liverpool to Manchester without a fresh supply of water. When a heavy load of merchandise is drawn, it is usual to take in water at the middle of the trip.

(93.) In reviewing all that has been stated, it will be perceived that the efficiency of the locomotive engines used on this railway is mainly owing to three circumstances: 1st, The unlimited power of draft in the furnace, by projecting the waste steam into the chimney; 2d, The unlimited abstraction of heat from the air passing from the furnace, by Mr. *Booth's* ingenious arrangement of tubes traversing the boiler; and, 3d, Keeping the cylinders warm, by immersing them in the chamber under the chimney.* There are many minor details which might be noticed with approbation, but these constitute the main features of the improvements, and should never, for a moment, be lost sight of by projectors of locomotive engines.

The successive introduction of improvements in the engines, some of which we have mentioned, has been accompanied by corresponding accessions to their practical power, and to the economy of fuel; and they have now arrived at a point which is as far beyond the former expectations of the most sanguine locomotive projectors, as it assuredly is short of the perfection of which these wonderful machines are still susceptible.

In the spring of the year 1832, I made several experiments on the Manchester railway, with a view to determine, in the actual state of the locomotive engines at that time, their powers with respect to the amount of load and the

* Mr. Robert Stephenson, whose experience and skill in the construction of locomotives attaches great importance to this condition. It has lately, however, been abandoned by some other engine makers, for the purpose of getting rid of the cranked axle which must accompany it.

economy of fuel. Since that time I am not aware that, in these respects, the engine has received any material improvement. The following are the particulars of three experiments thus made:—

I.

On Saturday, the 5th of May, the engine called the "Victory" took 20 wagons of merchandise, weighing gross 92 tons 19 cwt. 1 qr., together with the tender containing fuel and water, of the weight of which I have no account, from Liverpool to Manchester, (30 miles,) in 1 h. 34 min. 45 sec. The train stopped to take in water halfway, for 10 minutes, not included in the above-mentioned time. On the inclined plane rising 1 in 96, and extending 1½ mile, the engine was assisted by another engine called the "Samson," and the ascent was performed in 9 minutes. At starting, the fireplace was well filled with coke, and the coke supplied to the tender accurately weighed. On arriving at Manchester the fireplace was again filled, and the coke remaining in the tender weighed. The consumption was found to amount to 929 pounds net weight, being at the rate of one-third of a pound per ton per mile.

Speed on the level was 18 miles an hour; on a fall of 4 feet in a mile, 21½ miles an hour; fall of 6 feet in a mile, 25½ miles an hour; on the rise over Chatmoss, 8 feet in a mile 17⅗ miles an hour; on level ground sheltered from the wind, 20 miles an hour. The wind was moderate, but direct ahead. The working wheels slipped three times on Chatmoss, and the train was retarded from 2 to 3 minutes.

The engine, on this occasion, was not examined before or after the journey, but was presumed to be in good working order.

II.

On Tuesday, the 8th of May, the same engine performed the same journey, with 20 wagons, weighing gross 90 tons

7 cwt. 2 qrs., exclusive of the unascertained weight of the tender. The time of the journey was 1 h. 41 min. The consumption of coke 1040 lbs. net weight, estimated as before. Rate of speed;—

| | |
|---|---|
| Level - - - | $17\frac{5}{8}$ miles per hour. |
| Fall of 4 feet in a mile | 22 |
| ——— 6 - - - | $22\frac{1}{2}$ |
| Rise of 8 - - - | 15 |

On this occasion there was a high wind ahead on the quarter, and the connecting rod worked hot, owing to having been keyed too tight. On arriving at Manchester, I caused the cylinders to be opened, and found that the pistons were so loose, that the steam blew through the cylinders with great violence. By this cause, therefore, the machine was robbed of a part of its power during the journey; and this circumstance may explain the slight decrease in speed, and increase in the consumption of fuel, with a lighter load in this journey compared with that performed on the 5th of May.

The Victory weighs 8 tons 2 cwt., of which 5 tons 4 cwt. rest on the drawing wheels. The cylinders are 11 inches diameter, and 16 inches stroke; and the diameter of the drawing wheels is 5 feet.

## III.

On the 29th of May, the engine called the "Samson," (weighing 10 tons 2 cwt., with 14-inch cylinders, and 16-inch stroke; wheels 4 feet 6 inches diameter, both pairs being worked by the engine; steam 50 lbs. pressure, 130 tubes) was attached to 50 wagons, laden with merchandise; net weight about 150 tons; gross weight, including wagons, tender, &c., 223 tons 6 cwt. The engine with this load travelled from Liverpool to Manchester (30 miles) in 2 h. and 40 min., exclusive of delays upon the road for watering, &c., being at the rate of nearly 12 miles an hour. The speed varied according to the inclinations of the road. Upon a level, it was 12 miles an hour; upon a descent of 6 feet in a mile, it was 16 miles an hour: upon a rise of 8 feet in a mile, it was about 9 miles an hour. The weather was calm, the rails very wet; but the wheels did not slip, even in the slowest speed, except at starting, the rails being at that place soiled and greasy with the slime and dirt to which they are always exposed at the stations. The coke consumed in this journey, exclusive of what was raised in getting up the steam, was 1762 lbs., being at the rate of a quarter of a pound per ton per mile.

(94.) From the above experiments it appears that a locomotive engine, in good working order, with its full complement of load, is capable of transporting weights at an expense in fuel amounting to about four ounces of coke per ton per mile. The attendance required on the journey is that of an engine-man and a fire boy; the former being paid 1*s*. 6*d*. for each trip of 30 miles, and the latter 1*s*. In practice, however, we are to consider, that it rarely happens that the full complement of load can be sent with the engines; and when lesser loads are transported, the *proportionate expense* must, for obvious reasons, be greater.

The practical expenditure of fuel on the Liverpool and Manchester line may, perhaps, be fairly estimated at half a pound of coke per ton per mile.

(95.) Having explained the power and efficiency of these locomotive engines, it is now right to notice some of the defects under which they labour.

The great original cost, and the heavy expense of keeping the engines used on the railway in repair, have pressed severely on the resources of the undertaking. One of the best constructed of the later engines costs originally about 800*l*. It may be hoped that, by the excitement of competition, the facilities derived from practice, and from the manufacture of a greater number of engines of the same kind, some reduction of this cost may be effected. The original cost, however, is far from being the principal source of expense: the wear and tear of these machines, and the occasional fracture of those parts on which the greatest strain has been laid, have greatly exceeded what the directors had anticipated. Although this source of expense must be in part attributed to the engines not having yet attained that state of perfection, in the proportion and adjustment of their parts, of which they are susceptible, and to which experience alone can lead, yet there are some obvious defects which demand attention.

The heads of the boilers are flat, and formed of iron, similar to the material of the boilers themselves. The tubes which traverse the boiler were, until recently, copper, and so inserted into the flat head or ends as to be water-tight. When the boiler is heated, the tubes are found to expand in a greater degree than the other parts of the boiler; which frequently causes them either to be loosened at the extremities, so as to cause leakage, or to bend from want of room for expansion. The necessity of removing and refastening the tubes causes, therefore, a constant expense.

It will be recollected that the fireplace is situated at one end of the boiler, immediately below the mouths of the tubes: a powerful draught of air, passing through the fire, carries with it ashes and cinders, which are driven violently through the tubes, and especially the lower ones, situated near the fuel. These tubes are, by this means, subject to rapid wear, the cinders continually acting upon their interior surface. After a short time it becomes necessary to replace single tubes, according as they are found to be worn, by new ones; and it not unfrequently happens, when this is neglected, that tubes burst. After a certain length of time the engines require new tubing, which is done at the expense of about 70*l.*, allowing for the value of the old tubes. This wear of the tubes might possibly be avoided by constructing the fireplace in a lower position, so as to be more removed from their mouths; or, still more effectually, by interposing a casing of metal, which might be filled with water, between the fireplace and those tubes which are the most exposed to the cinders and ashes. The unequal expansion of the tubes and boilers appears to be an incurable defect, if the present form of the engine be retained. If the fireplace and chimney could be placed at the same end of the boiler, so that the tubes might be recurved, the unequal expansion would then produce no injurious effect; but it would be difficult to clean the tubes if they were exposed, as they are at present, to the cinders. The next source of expense arises from the wear of the boiler-head, which is exposed to the action of the fire. These require constant patching and frequent renewal.

A considerable improvement has lately been introduced into the method of tubing, by substituting brass for copper tubes. We are not aware that the cause of this improvement has been discovered; but it is certain, whatever be the cause, that brass tubes are subject to considerably slower wear than copper.

It has been said by some whose opinions are adverse to the advantage of railways, but more especially to the particular species of locomotive engines now under consideration, that the repairs of one of these engines cost so great a sum as 1500*l.* per annum, and that the directors now think of abandoning them, or adopting either stationary engines or horse power. As to the first of these statements I must observe, that the expense of repairs of such machines should never be computed in reference to *time*, but rather to the work done, or the distance travelled over. I have ascertained that engines frequently travel a distance of from 25,000 to 30,000 miles before they require new tubing. During that work, however, single tubes are, of course, occasionally renewed, and other repairs are made, the expense of which may safely be stated as under the original cost of the engine. The second statement, that the company contemplate substituting stationary engines, or horses, for locomotives, is altogether at variance with the truth. Whatever improvements may be contemplated in locomotives, the directors assuredly have not the slightest intention of going back in the progress of improvement, in the manner just mentioned.

The expense of locomotive power having so far exceeded what was anticipated at the commencement of the undertaking, it was thought advisable, about the beginning of the year 1834, to institute an inquiry into the causes which produced the discrepancy between the estimated and actual expenses, with a view to the discovery of some practical means by which they could be reduced. The directors of the company, for this purpose, appointed a sub-committee of their own body, assisted by Mr. *Booth*, their treasurer, to inquire and report respecting the causes of the amount of this item of their expenditure, and to ascertain whether any and what measures could be devised for the attainment of greater economy. A very able and satisfactory report was made by this committee, or, to speak more correctly, by Mr. *Booth.*

It appears that, previous to the establishment of the railway, Messrs. *Walker* and *Rastrick*, engineers, were employed by the company to visit various places where steam power was applied on railways, for the purpose of forming an estimate of the probable expense of working the railway by locomotive and by fixed power. These engineers recommended the adoption of locomotive power, and their estimate was, that the transport might be effected at the rate of .278 of a penny, or very little more than a farthing per ton per mile. In the year 1833, five years after this investigation took place, it was found that the actual cost was .625 of a penny, or something more than a halfpenny per ton per mile, being considerably above double the estimated rate. Mr. *Booth* very properly directed his inquiries to ascertain the cause of this discrepancy, by comparing the various circumstances assumed by Messrs. *Walker* and *Rastrick*, in making their estimate, with those under which the transport was actually effected. The first point of difference which he observed was the *speed* of transport: the estimate was founded on an assumed speed of ten miles an hour, and it was stated that a four-fold speed would require an addition of 50 per cent. to the power, without taking into account wear and tear. Now the actual speed of transport being double the speed assumed in the statement, Mr. *Booth* holds it to be necessary to add 25 per cent. on that score.

The next point of difference is in the amount of the loads: the estimate is founded upon the assumption, that every engine shall start with its full complement of load, and that with this it shall go the whole distance. "The facts, however, are," says Mr. *Booth*, "that, instead of a *full load* of profitable carriage *from* Manchester, about half the wagons *come back empty*, and, instead of the tonnage being conveyed the whole way, many thousand tons are conveyed only half the way; also, instead of the daily work being uniform, it is extremely fluctuating." It is further remarked, that in order to accomplish the transport of goods from the branches and from intermediate places, engines are despatched several times a day, from both ends of the line, *to clear the road;* the object of this arrangement being rather to lay the foundation of a beneficial intercourse in future, than with a view to any immediate profit. Mr. *Booth* makes a rough estimate of the disadvantages arising from these circumstances by stating them at 33 per cent. in addition to the original estimate.

The next point of difference is the fuel. In the original estimate *coal* is assumed as the fuel, and it is taken at the price of five shillings and tenpence per ton: now the act of Parliament forbids the use of coal which would produce smoke; the company have, therefore, been obliged to use *coke*, at seventeen shillings and sixpence a ton. Taking coke, then, to be equivalent to coal, ton for ton, this would add .162 to the original estimate.

These several discrepancies being allowed for, and a proportional amount being added to the original estimate, the amount would be raised to .601 of a penny per ton per mile, which is within one-fortieth of a penny of the actual cost. This difference is considered to be sufficiently accounted for by the wear and tear produced by the very rapid motion, more especially when it is considered that many of the engines were constructed before the engineer was aware of the great speed that would be required.

"What then," says Mr. *Booth*, in the Report already alluded to, "is the result of these opposite and mutually counteracting circumstances? and what is the present position of the company in respect of their moving power? Simply, that they are still in a course of experiment, to ascertain practically the best construction, and the most durable materials, for engines required to transport greater weights, and at greater velocities, than had, till very recently, been considered possible; and which, a few years ago, it had not entered into the imagination of the most daring and sanguine inventor to conceive: and, farther, that these experiments have necessarily been made, not with the calm deliberation and quiet pace which a salutary caution recommends,—making good each step in the progress of discovery before advancing another stage,—but amid the bustle and responsibilities of a large and increasing traffic; the directors being altogether ignorant of the time each engine would last before it would be laid up as inefficient, but compelled to have engines, whether good or bad; being aware of various defects and imperfections, which it was impossible at the time to remedy, yet obliged to keep the machines in motion, under all the disadvantages of heavy repairs, constantly going on during the night, in order that the requisite number of engines might be ready for the morning's work. Neither is this great experiment yet complete; it is still going forward. But the most prominent difficulties have been in a great measure surmounted ; and your committee conceive, that they are warranted in expecting, that the expenditure in this department will, ere long, be materially reduced, more especially when they consider the relative performances of the engines at the *present time* compared with what it was two years ago."

In the half year ending 31st December, 1831, the six best engines performed as follows :—

| | Miles. |
|---|---|
| Planet - - - | 9,986 |
| Mercury - - - - | 11,040 |
| Jupiter - - - | 11,618 |
| Saturn - - - - | 11,786 |
| Venus - - - | 12,850 |
| Etna - - - - | 8,764 |
| Making in all - - | 66,044 |

In the half year ending 31st December, 1833, the six best engines performed as follows :—

| | Miles. |
|---|---|
| Jupiter - - - | 16,572 |
| Saturn - - - | 18,678 |
| Sun - - - | 15,552 |
| Etna - - - - | 17,763 |
| Ajax - - - | 11,678 |
| Firefly - - - - | 15,608 |
| Making in all - - | 95,851 |

(96.) The advantages derivable from railroads are greatly abridged by the difficulty arising from those changes of level to which all roads are necessarily liable ; but in the case of railroads, from causes peculiar to themselves, these changes of level occasion great inconvenience. To explain the nature of these difficulties, it will be necessary to consider the relative proportion which must subsist between the power of traction on a level and on an inclined plane. On a level railroad the force of traction necessary to propel any load, placed on wheel carriages of the construction now commonly used, may perhaps be estimated at 7½ pounds,* for every ton gross in the load; that is to say, if a load of one ton gross were placed upon wheel carriages upon a level railroad, the traces of horses drawing it would be stretched with a force equivalent to 7½ pounds. If the load amounted to two or three tons, the tension of the traces would be increased to 15 or 22½ pounds, and so on. The necessity of this force of traction, arising from the want of perfect smoothness in the road, and from the friction of the wheels and axles of the carriages, must be the same whether the road be level or inclined; and consequently, in ascending an inclined plane, the same force of traction will be necessary in addition to that which arises from the tendency of the load to fall down the plane. This latter tendency is always in the proportion of the elevation of the plane to its length; that is to say, a plane which rises 1 foot in 100 will give a weight of 100 tons a tendency to fall down the plane amounting to 1 ton, and would therefore add 1 ton to the force of traction necessary for such a load on a level.

Now since 7½ pounds is very nearly the 300th part of a ton, it follows that if an inclination upon a railroad rises at the rate of 1 foot in 300, or, what is the same, 17½ feet in a mile, such an acclivity will add 7½ pounds per ton to the force of traction. This acclivity therefore would require a force of traction twice as great as a level. In like manner a rise of 35 feet in a mile would require three times the force

* The estimate commonly adopted by engineers at present is 9 pounds per ton. I have no doubt, however, that this is too high. I am now (November, 1835) engaged in an extensive course of experiments on different railways, with a view to determine with precision this and other points connected with the full developement of their theory; and I have reason to believe, from the observations I have already made, that even 7½ pounds per ton is above the average force of traction upon the level.

of traction of a level, $52\frac{1}{2}$ feet in a mile four times that force, and so on. In fact, for every seven feet in a mile which an acclivity rises, 3 pounds per ton will be added to the force of traction. If we would then ascertain the power necessary to pull a load up any given acclivity upon a railroad, we must first take $7\frac{1}{2}$ pounds as the force necessary to overcome the common resistance of the road, and then add 3 pounds for every 7 feet which the acclivity rises per mile. For example, suppose an acclivity to rise at the rate of 70 feet in a mile, the force of traction necessary to draw a ton up it would be thus calculated:—

| | |
|---|---|
| Friction - - - - | - $7\frac{1}{2}$ lbs. |
| 70 feet = 10 times 3 lbs. | - 30 |
| Total force - - | - $37\frac{1}{2}$ |

It will be apparent, therefore, that if a railroad undulates by inclined planes, even of the most moderate inclinations, the propelling power to be used upon it must be of such a nature as to be capable of increasing its intensity in a great degree, according to the elevation of the planes which it has to encounter. A plane which rises $52\frac{1}{2}$ feet per mile presents to the eye scarcely the appearance of an ascent, and yet requires the power of traction to be increased in a fourfold proportion.

It is the property of animal power, that within certain limits its energy can be put forth at will, according to the exigency of the occasion; but the intensity of mechanical power, in the instance now considered, cannot so conveniently be varied, except indeed within narrow limits.

In the application of locomotive engines upon railways the difficulty arising from inclined planes has been attempted to be surmounted by several methods, which we shall now explain.

1. Upon arriving at the foot of the plane the load is divided, and the engine carries it up in several successive trips, descending the plane unloaded after each trip. The objec-

tion to this method is the delay which it occasions—a circumstance which is incompatible with a large transport of passengers. From what has been stated, it would be necessary, when the engine is fully loaded on a level, to divide its load into four parts, to be successively carried up when the incline rises 52 feet per mile. This method has been practised in the transport of merchandise occasionally, when heavy loads were carried on the Liverpool and Manchester line, upon the Rainhill incline.

2. A subsidiary or assistant locomotive engine may be kept in constant readiness at the foot of each incline, for the purpose of aiding the different trains, as they arrive, in ascending. The objection to this mode is the cost of keeping such an engine with its boiler continually prepared, and its steam up. It would be necessary to keep its fire continually lighted, whether employed or not; otherwise, when the train would arrive at the foot of the incline, it should wait until the subsidiary engine was prepared for work. In cases where trains would start and arrive at stated times, this objection, however, would have less force. This method is at present generally adopted on the Liverpool and Manchester line. This method, however, cannot be profitably applied to planes of any considerable length.

3. A fixed steam engine may be erected on the crest of the incline, so as to communicate by ropes with the train at the foot. Such an engine would be capable of drawing up one or two trains together, with their locomotives, according as they would arrive, and no delay need be occasioned. This method requires that the fixed engine should be kept constantly prepared for work, and the steam continually up in the boiler. This expedient is scarcely compatible with a large transit of passengers, except at the terminus of a line.

4. In working on the level, the communication between the boiler and the cylinder in the locomotives may be so estrained by partially closing the throttle valve, as to cause he pressure upon the piston to be less in a considerable de-

gree than the pressure of steam in the boiler. If under such circumstances a sufficient pressure upon the piston can be obtained to draw the load on the level, the throttle valve may be opened on approaching the inclined plane, so as to throw on the piston a pressure increased in the same proportion as the previous pressure in the boiler was greater than that upon the piston. If the fire be sufficiently active to keep up the supply of steam in this manner during the ascent, and if the rise be not greater in proportion than the power thus obtained, the locomotive will draw the load up the incline without further assistance. It is, however, to be observed, that in this case the load upon the engine must be less than the amount which the adhesion of its working wheels with the railroad is capable of drawing; for this adhesion must be adequate to the traction of the same load up the incline, otherwise, whatever increase of power might be obtained by opening the throttle valve, the drawing wheels would revolve without causing the load to advance. This method has been generally practised upon the Liverpool and Manchester line in the transport of passengers; and, indeed, it is the only method yet discovered, which is consistent with the expedition necessary for that species of traffic. The objections to this method are, the necessity of maintaining a much higher pressure in the boiler than is sufficient for the purposes of the load upon more level parts of the line.

In the practice of this method considerable aid may be derived also by suspending the supply of feeding water during the ascent. It will be recollected that a reservoir of cold water is placed in the tender which follows the engine, and that the water is driven from this reservoir into the boiler by a forcing pump, which is worked by the engine itself. This pump is so constructed that it will supply as much cold water as is equal to the evaporation, so as to maintain constantly the same quantity of water in the boiler. But it is evident, on the other hand, that the supply of this water has a tendency to check the rate of evaporation, since in being raised to the temperature of the water with which it mixes, it must absorb a considerable portion of the heat supplied by the fire. With a view to accelerate the production of steam, therefore, in ascending the inclines, the engine-man may suspend the action of the forcing pump, and thereby stop the supply of cold water to the boiler; the evaporation will go on with increased rapidity, and the exhaustion of water produced by it will be repaid by the forcing pump on the next level, or still more effectually on the next descending incline. Indeed, the feeding pump may be made to act in descending an incline if necessary, when the action of the engine itself is suspended, and when the train descends by its own gravity, in which case it will perform the part of a brake upon the descending train.

This method, on railroads intended for passengers, may be successfully applied on inclines which do not exceed 18 feet in a mile; and, with a sacrifice of the expense of locomotive power, inclines so steep as 36 feet in a mile may be worked in this manner. As, however, the sacrifice is considerable, it will, perhaps, be always better to work the more steep inclines by assistant engines.

5. The mechanical connexion between the piston of the cylinder and the points of contact of the working wheels with the road may be so altered, upon arriving at the incline, as to give the piston a greater power over the working wheels. This may be done in an infinite variety of ways, but hitherto no method has been suggested sufficiently simple to be applicable in practice; and even were any means suggested which would accomplish this, unless the intensity of the impelling power were at the same time increased, it would necessarily follow that the speed of the motion would be diminished in exactly the same proportion as the power of the piston over the working wheels would be increased. Thus, on the inclined plane, which rises 55 feet per mile, upon the Liverpool line, the speed would be diminished to nearly one-fourth of its amount upon the level.

Whatever be the method adopted to surmount inclined planes upon a railway, inconvenience attends the descent upon them. The motion down the incline by the force of gravity is accelerated; and if the train be not retarded, a descent of any considerable length, even at a small elevation, would produce a velocity which would be attended with great danger. The shoe used to retard the descent down hills on turnpike roads cannot be used upon railroads, and the application of brakes to the faces of the wheels is likewise attended with some uncertainty. The friction produced by the rapid motion of the wheel sometimes sets fire to wood, and iron would be inadmissible. The action of the steam on the piston may be reversed, so as to oppose the motion of the wheels; but even this is attended with peculiar difficulty.

From all that has been stated, it will be apparent that, with our present knowledge, considerable inclines are fatal to the profitable performance on a railway, and even small inclinations are attended with great inconvenience.*

(97.) To obtain from the locomotive steam engines now used on the railway the most powerful effects, it is necessary that the load placed on each engine should be very considerable. It is not possible, with our present knowledge, to construct and work three locomotive engines of this kind, each drawing a load of 30 tons, at the same expense and with the same effect as one locomotive engine drawing 90 tons. Hence arises what must appear an inconvenience and difficulty in applying these engines to one of the most profitable species of transport—the transport of passengers. It is impracticable, even between places of the most considerable

* A contrivance might be applied in changes of level in railroads somewhat similar to locks in a canal. The train might be rolled upon a platform which might be raised by machinery; and thus at the change of level there would be as it were *steps* from one level to another, up which the loads would be lifted by any power applied to work the machinery. The advantage in this case would be, that the trains might be adapted to work always upon a level.

intercourse to obtain loads of passengers sufficiently great at each trip to maintain such an engine working on a railway.* The difficulty of collecting so considerable a number of persons, at any stated hour, to perform the journey, is obvious; and therefore, the only method of removing the inconvenience is *to cause the same engine which transports passengers also to transport goods*, so that the goods may make up the requisite supplement to the load of passengers. In this way, provided the traffic in goods be sufficient, such engines may start with their full complement of load, whatever be the number of passengers.

(98.) In comparing the extent of capital, and the annual expenditure of the Liverpool and Manchester line, and adopting it as a modulus in estimating the expenses of similar undertakings projected elsewhere, there are several circumstances to which it is important to attend. I have already observed on the large waste of capital in the item of locomotive engines which ought to be regarded as little more than experimental machines, leading to a rapid succession of improvements. Most of these engines are still in good working order, but have been abandoned for the reasons already assigned. Other companies will, of course, profit by the experience which has thus been purchased at a high price by the Liverpool Company. This advantage in favour of future companies will go on increasing until such companies have their works completed.

A large portion of the current expense of a line of railway is independent of its length; and is little less for the line connecting Liverpool and Manchester, than it would

* On the occasion of races held at Newton, a place about fifteen miles from Liverpool, two engines were sent, with trains of carriages, to take back to Liverpool the visiters to the races. Some accident prevented one of the engines from working on the occasion, and both trains were attached to the same engine: 800 persons were on this occasion drawn by the single engine to Liverpool in the space of about an hour.

be for a line connecting Birmingham with Liverpool or London.

The establishments of resident engineers, coach and wagon yards, &c. at the extremities of the line, would be little increased by a very great increase in the length of the railway; and the same observation will apply to other heads of expenditure.

It has been the practice of the canal companies between Liverpool and Manchester to warehouse the goods transported between these towns, without any additional charge beyond the price of transport. The Railway Company, in competing with the canals, were, of course, obliged to offer like advantages: this compelled them to invest a considerable amount of capital in the building of extensive warehouses, and to incur the annual expense of porterage, salaries, &c. connected with the maintenance of such storage. In a longer line of railway such expenses (if necessary at all) would not be proportionally increased.

(99.) The comparison of steam-transport with the transport by horses, even when working on a railway, exhibits the advantage of this new power in a most striking point of view. To comprehend these advantages fully, it will be necessary to consider the manner in which animal power is expended as a means of transport. The portion of the strength of a horse available for the purpose of a load depends on the speed of a horse's motion. To this speed there is a certain limit, at which the whole power of the horse will be necessary to move his own body, and at which, therefore, he is incapable of carrying any load; and, on the other hand, there is a certain load which the horse is barely able to support, but incapable of moving with any useful speed. Between these two limits there is a certain rate of motion at which the useful effect of the animal is greatest. In horses of the heavier class, this rate of motion may be taken on the average as that of 2 miles an hour; and in the lighter description of horses, $2\frac{1}{2}$ miles an hour. Beyond this speed, the load which they are capable of transporting diminishes in a very rapid ratio as the speed increases: thus, if 121 express the load which a horse is able to transport a given distance in a day, working at the rate of four miles an hour, the same horse will not be able to transport more than the load expressed by 64, *the same distance*, at 7 miles an hour; and, at 10 miles an hour, the load which he can transport will be reduced to 25. The most advantageous speed at which a horse can work being 2 miles an hour, it is found that, at this rate, working for 10 hours daily, he can transport 12 tons, on a level railway, a distance of 20 miles; so that the whole effect of a day's work may be expressed by 240 tons carried 1 mile.

But this rate of transport is inapplicable to the purposes of travelling; and therefore it becomes necessary, when horses are the moving power, to have carriages for passengers distinct from those intended for the conveyance of goods; so that the goods may be conveyed at that rate of speed at which the whole effect of the horse will be the greatest possible; while the passengers are conveyed at that speed which, whatever the cost, is indispensably necessary. The weight of an ordinary mail-coach is about two tons; and, on a tolerably level turnpike road, it travels at the rate of 10 miles an hour. At this rate, the number of horses necessary to keep it constantly at work, including the spare horses indispensably necessary to be kept at the several stages, is computed at the rate of a horse per mile. Assuming the distance between London and Birmingham at 100 miles, a mail-coach running between these two places would require 100 horses; making the journey to and from Birmingham daily. The performance, therefore, of a horse working at this rate may be estimated at 2 tons carried 2 miles per day, or 4 tons carried 1 mile in a day. The force of traction on a good turnpike road is at least 20 times its amount on a level railroad. It therefore follows, that the performance of a horse on a railroad will be 20 times the amount of its performance on a common road under similar circumstances. We may, therefore, take the performance of a horse working at 10 miles an hour, on a level railroad, at 80 tons conveyed 1 mile daily.

The best locomotive engines used on the Liverpool railway are capable of transporting 150 tons on a level railroad at the same rate; and, allowing the same time for stoppage, its work per day would be 150 tons conveyed 200 miles, or 30,000 tons conveyed 1 mile; from which it follows, that the performance of one locomotive engine of this kind is equivalent to that of 7500 horses working on a good turnpike road, or to 375 horses working on a railway. The consumption of fuel requisite for this performance, with the most improved engines used at present on the Manchester and Liverpool line, would be at the rate of eight* ounces of coke per ton per mile, including the waste of fuel incurred by the stoppages. Thus the daily consumption of fuel, under such circumstances, would amount to 15000 lbs. of coke; and 2 lbs. of coke daily would perform the work of one horse on a good turnpike road; and 40 lbs. of coke daily would perform the work of one horse on a railway.

In this comparison, the engine is taken at its most advantageous speed, while horse-power is taken at its least advantageous speed, if regard be only had to the total quantity of weight transported to a given distance. But, in the case above alluded to, speed is an indispensable element; and steam, therefore, possesses this great advantage over horse-power, that *its most advantageous speed is that which is at once adapted to all the purposes of transport, whether of passengers or of goods.*

(100.) The effects of steam compared with horse-power, at lower rates of motion, will exhibit the advantages of the former, though in a less striking degree. An eight-horse

* In an experimental trip with a heavy train at 12 miles an hour, I found the consumption of coke to be only *four* ounces per ton per hour. I believe, however, the practical consumption in ordinary work to be very nearly eight ounces.

wagon commonly weighs 8 tons, and travels at the rate of $2\frac{1}{2}$ miles an hour. Strong horses working in this way can travel 8 hours daily ; thus each horse performs 20 miles a day. The performance, therefore, of each horse may be taken as equivalent to 20 tons transported 1 mile ; and his performance on a railway being 20 times this amount, may be taken as equivalent to 400 tons transported 1 mile a day. The performance of a horse working in this manner is, therefore, 5 times the performance of a horse working at 10 miles an hour; the latter effecting only the performance of 4 tons transported 1 mile per day on a good turnpike road, or 80 tons on a railway. We shall hence obtain the proportion of the performance of horses working in wagons to that of a locomotive steam engine. Since 2 lbs. of coke are equivalent to the daily performance of a horse in a mail-coach, and 40 lbs. on a railway, at 10 miles an hour, it follows that 10 lbs. will be equivalent to the performance of a horse on a turnpike road, and 200 lbs. on a railway, at $2\frac{1}{2}$ miles an hour. Since a locomotive engine can perform the daily work of 7500 mail-coach horses, it follows that it performs the work of 1500 wagon horses.

These results must be understood to be subject to modifications in particular cases, and to be only average calculations. Different steam-engines, as well as different horses, varying in their performance to a considerable extent ; and the roads on which horses work being in different states of perfection, and subject to different declivities, the performance must vary accordingly.

In the practical comparison, also, of the results of so powerful an agent as steam applied on railways, with so slight a power as that of horses on common roads, it must be considered that the great subdivision of load, and frequent times of starting, operate in favour of the performance of horses ; inasmuch as it would oftener occur that engines capable of transporting enormous weights would start with loads inferior to their power, than would happen in the application of horse-power, where small loads may start at short intervals

This, in fact, constitutes a practical difficulty in the applica tion of steam engines on railroads; and will, perhaps, for the present, limit their application to lines connecting places of great intercourse.

The most striking effect of steam power, applied on a railroad, is the extreme speed of transport which is attained by it; and it is the more remarkable, as this advantage never was foreseen before experience proved it. When the Liverpool and Manchester line was projected, the transport of heavy goods was the object chiefly contemplated; and although an intercourse in passengers was expected, it was not foreseen that this would be the greatest source of revenue to the proprietors. The calculations of future projectors will, therefore, be materially altered, and a great intercourse in passengers will be regarded as a necessary condition for the prosperity of such an undertaking.

If this advantage of speed be taken into account, horse-power can scarcely admit of any comparison whatever with steam-power on a railway. In the experiments which I have already detailed, it appears that a steam engine is capable of drawing 90 tons at the rate of about 20 miles an hour, and that it could transport that weight twice between Liverpool and Manchester in about 3 hours. Two hundred and seventy horses working in wagons would be necessary to transport the same load the same distance in a day. It may be objected, that this was an experiment performed under favourable circumstances, and that assistance was obtained at the difficult point of the inclined plane. In the ordinary performance, however, of the engines drawing merchandise, where great speed is not attempted, the rate of motion is not less than 15 miles an hour. In the trains which draw passengers, the chief difficulty of maintaining a great speed arises from the stoppages on the road to take up and let down passengers. There are two classes of carriages at present used: the first class stops but once, at a point halfway between Liverpool and Manchester, for the space of a few minutes. This class performs the 30 miles in an hour and a half, and sometimes in 1 hour and 10 minutes. On the level part of the road its common rate of motion is 27 miles an hour; and I have occasionally marked its rate, and found it above 30 miles an hour.

But these, which are velocities obtained in the regular working of the engines for the transport of passengers and goods, are considerably inferior to the power of the present locomotives with respect to speed. I have made some experimental trips, in which more limited loads were placed upon the engines, by which I have ascertained that very considerably increased rates of motion are quite practicable. In one experiment I placed a carriage containing 36 persons upon an engine, with which I succeeded in obtaining the velocity of about 48 miles an hour, and I believe that an engine loaded only with its own tender has moved over 15 miles in 15 minutes.

It will then perhaps be asked, if the engines possess these great capabilities of speed, why they have not been brought into practical operation on the railroad, where, on the other hand, the average speed when actually in motion does not exceed 25 miles an hour? In answer to this it may be stated, that the distance of 30 miles between Liverpool and Manchester is performed in an hour and a half, and that 10 trains of passengers pass daily between these places: the mail, also, is transmitted three times a day between them. It is obvious that any greater speed than this, in so short a distance, would be quite needless. When, however, more extended lines of road shall be completed, the circumstances will be otherwise, and the despatch of mails especially will demand attention. Full trains of passengers, commonly transported upon the Manchester railroad, weigh about 50 tons gross; with a lighter load, a lighter and more expeditious engine might be used. The expense of transport with such an engine would of course be increased; but for this the increased expedition there would be ample compensation

When, therefore, London shall have been connected with Liverpool by a line of railroad through Birmingham, the commercial interest of these places will naturally direct attention to the greatest possible expedition of intercommunication. For the transmission of mails, doubtless, peculiar engines will be built, adapted to lighter loads and greater speed. With such engines, the mails, with a limited number of passengers, will be despatched; and, apart from any possible improvement which the engines may hereafter receive, and looking only at their present capabilities, I cannot hesitate to express my conviction that such a load may be transported at the rate of above 60 miles an hour. If we may indulge in expectations of what the probable improvements of locomotive steam engines may effect, I do not think that even double that speed is beyond the limits of mechanical probability. On the completion of the line of road from the metropolis to Liverpool we may, therefore, expect to witness the transport of mails and passengers in the short space of three hours. There will probably be about three posts a day between these and intermediate places.

The great extension which the application of steam to the purpose of inland transport is about to receive from the numerous railroads which are already in progress, and from a still greater number of others which are hourly projected, impart to these subjects of inquiry considerable interest. Neither the wisdom of the philosopher, nor the skill of the statistician, nor the foresight of the statesman is sufficient to determine the important consequences by which the realization of these schemes must affect the progress of the human race. How much the spread of civilization, the diffusion of knowledge, the cultivation of taste, and the refinement of habits and manners depend upon the easy and rapid intermixture of the constituent elements of society, it is needless to point out. While population exists in detached and independent masses, incapable of transfusion among each other, their dormant affinities are never called into action, and the most precious qualities of each are never imparted to the other. Like solids in physics, they are slow to form combinations; but when the quality of fluidity has been imparted to them, when their constituent atoms are loosened by fusion, and the particles of each flow freely through and among those of the other, then the affinities are awakened, new combinations are formed, a mutual interchange of qualities takes place, and compounds of value far exceeding those of the original elements are produced. Extreme facility of intercourse is the fluidity and fusion of the social masses, from whence such an activity of the affinities results, and from whence such an inestimable interchange of precious qualities must follow. We have, accordingly, observed, that the advancement in civilization and the promotion of intercourse between distant masses of people have ever gone on with contemporaneous progress, each appearing occasionally to be the cause or the consequence of the other. Hence it is that the urban population is ever in advance of the rural in its intellectual character. But, without sacrificing the peculiar advantages of either, the benefits of intercourse may be extended to both, by the extraordinary facilities which must be the consequence of the locomotive projects now in progress. By the great line of railroad which is in progres from London to Birmingham, the time and expense of passing between these places will probably be halved, and the quantity of intercourse at least quadrupled, if we consider only the direct transit between the terminal points of the line; but if the innumerable tributary streams which will flow from every adjacent point be considered, we have no analogies on which to build a calculation of the enormous increase of intercommunication which must ensue.

Perishable vegetable productions necessary for the wants of towns must at present be raised in their immediate suburbs; these, however, where they can be transported with a perfectly smooth motion at the rate of twenty miles an hour, will be supplied by the agricultural labourer of more distant points. The population engaged in towns, no longer limited to their narrow streets, and piled story over story in confined habitations, will be free to reside at distances which would now place them far beyond reach of their daily occupations. The salubrity of cities and towns will thus be increased by spreading the population over a larger extent of surface, without incurring the inconvenience of distance. Thus the advantages of the country will be conferred upon the town, and the refinement and civilization of the town will spread their benefits among the rural population.*

(101.) The quantity of canal property in these countries gives considerable interest to every inquiry which has for its object the relative advantage of this mode of transport, compared with that of railways, whether worked by horses or by steam power; and this interest has been greatly increased by the recent extension of railway projects. This is a subject which I shall have occasion, in another work, to examine in all its details; and, therefore, in this place I shall advert to it but very briefly.

When a floating body is moved on a liquid, it will suffer a resistance, which will depend partly upon the transverse section of the part immersed, and partly on the speed with which it is moved. It is evident that the quantity of the liquid which it must drive before it will depend upon that transverse section, and the velocity with which it will impel the liquid will depend upon its own speed. Now, so long as the depth of its immersion remains the same, it is demonstrable that the resistance will increase in proportion to the square of the speed; that is, with a double velocity there will be a fourfold resistance, with a triple velocity a ninefold resistance, and so on. Again, if the part immersed should be increased or diminished by any cause, the resistance, on

* Some of the preceding observations appeared in an article contributed by me to the British and Foreign Review.

that account alone, will be increased or diminished in the same proportion.

From these circumstances it will be apparent that a vessel floating on water, if moved with a certain speed, will require four times the impelling force to carry it forward with double the speed, unless the depth of its immersion be diminished as its speed is increased.

Some experiments which have been made upon canals with boats of a peculiar construction, drawn by horses, have led to the unexpected conclusion, that, after a certain speed has been attained, the resistance, instead of being increased, has been diminished. This fact is not at variance with the law of resistance already explained. The cause of the phenomenon is found in the fact, that when the velocity has attained a certain point, the boat gradually rises out of the water; so that, in fact, the immersed part is diminished. The two conditions, therefore, which determine the resistance, thus modify each other: while the resistance is, on the one hand, increased in proportion to the square of the speed, it is, on the other hand, diminished in proportion to the diminution of the transverse section of the immersed part of the vessel. It would appear that, at a certain velocity, these two effects neutralize each other; and, probably, at higher velocities the immersed part may be so much diminished as to diminish the resistance in a greater degree than it is increased by the speed, and thus actually to diminish the power of traction.

It is known that boats are worked on some of the Scottish canals, and also on the canal which connects Kendal with Preston, by which passengers are transported at the rate of about ten miles an hour, exclusive of the stoppages at the locks, &c. The power of horses, exerted in this way, is, of course, exerted more economically than they could be worked at the same speed on common roads; and, probably, it is as economical as they would be worked by railroad. It is, probably, more economical than the transport of passengers by steam upon railroads; but the speed is considerably less, nor, from the nature of the impelling power, is it possible that it can be increased.

There is reason to suppose that a like effect takes place with steam vessels. Upon increasing the power of the engines in some of the Post-office steam packets, it has been found, that, while the time of performing the same voyage is diminished, the consumption of fuel is also diminished. Now since the consumption of fuel is in the direct ratio of the moving power, and the latter in the direct ratio of the resistance, it follows that the resistance must in this case be likewise diminished.

(102.) When a very slow rate of travelling is considered, the useful effects of horse-power applied on canals is somewhat greater than the effect of the same power applied on railways; but at all speeds above three miles an hour, the effect on railways is greater; and when the speed is considerable, the canal becomes wholly inapplicable, while the railway loses none of its advantages. At three miles an hour, the performance of a horse on a canal and a railway is in the proportion of four to three to the advantage of the canal; but at four miles an hour his performance on a railway has the advantage in very nearly the same proportion. At six miles an hour, a horse will perform three times more work on a railway than on a canal. At eight miles an hour, he will perform nearly five times more work.

But the circumstance which, so far as respects passengers, must give railways, as compared with canals, an advantage which cannot be considered as less than fatal to the latter, is the fact, that the great speed and cheapness of transit attainable upon a railway by the aid of steam-power will always secure to such lines not only a monopoly of the travelling, but will increase the actual amount of that source of profit in an enormous proportion, as has been already made manifest between Liverpool and Manchester. Before the opening of the railway there were about twenty-five coaches daily running between Liverpool and Manchester. If we assume these coaches on the average to take ten persons at each trip, it will follow that the number of persons passing daily between these towns was about 500. Let us, then, assume that 3000 persons passed weekly. This gives in six months 78,000. In the six months which terminated on the 31st of December, 1831, the number of passengers between the same towns, exclusive of any taken up on the road, was 256,321; and if some allowance be made for those taken up on the road, the number may be fairly stated at 300,000. At present there is but one coach on the road between Liverpool and Manchester; and it follows, therefore, that, besides taking the monopoly of the transit in travellers, the actual number has been already increased in a fourfold proportion.

The monopoly of the transit of passengers thus secured to the line of communication by railroad will always yield so large a profit as to enable merchandise to be carried at a comparatively low rate.

In light goods, which requires despatch, it is obvious that the railroad will always command the preference; and the question between that mode of communication and canals is circumscribed to the transit of those classes of heavy goods in which even a small saving in the cost of transport is a greater object than despatch.

(103.) The first effect which the Liverpool railroad produced on the Liverpool and Manchester canals was a fall in the price of transport; and at this time, I believe, the cost of transport per ton on the railroads and on the canals is the same. It will, therefore, be naturally asked, this being the case, why the greater speed and certainty of the railroad does not in every instance give it the preference, and altogether deprive the canals of transport? This effect, however, is prevented by several local and accidental causes, as well as by direct influence and individual interest. A large portion of the commercial and manufacturing population of Liverpool and Manchester have property invested in the canals, and are deeply interested to sustain them in opposition to the railway. Such persons will give the preference to the canals in their own business, and will induce those over whom they have influence to do so in every case where speed of transport is not absolutely indispensable.

Besides these circumstances, the canal communicates immediately with the shipping at Liverpool, and it ramifies in various directions through Manchester, washing the walls of many of the warehouses and factories for which the goods transported are destined. The merchandise is thus transferred from the shipping to the boat, and brought directly to the door of its owner, or *vice versâ*. If transported by the railway, on the other hand, it must be carried to the station at one extremity; and, when transported to the station at the other, it has still to be carried to its destination in different parts of the town.

These circumstances will sufficiently explain why the canals still retain, and may probably continue to retain, a share of the traffic between these great marts.

# CHAPTER XI.

## LOCOMOTIVE ENGINES ON TURNPIKE ROADS.

**Railways and Turnpike Roads compared.—Mr. Gurney's Inventions.—His Locomotive Steam Engine.—Its Performances.—Prejudices and Errors.—Committee of the House of Commons.—Convenience and Safety of Steam Carriages.—Hancock's Steam Carriage.—Mr. N. Ogle.—Trevithick's Invention.—Proceedings against Steam Carriages.—Turnpike Bills.—Steam Carriage between Gloucester and Cheltenham.—Its Discontinuance.—Report of the Committee of the Commons.—Present State and Prospects of Steam Carriages.**

(104.) We have hitherto confined our observations to steam-power as a means of transport applied on railways, but modern speculation has not stopped here. Several attempts have been made, and some of them attended with considerable success, to work steam carriages on turnpike roads. The practicability of this project has been hitherto generally considered to be very questionable; but if we carry back our view to the various epochs in the history of the invention of the steam engine, we shall find the same doubt, and the same difficulty, started at almost every important step in its progress. In comparing the effect of a turnpike road with that of a railway, there are two circumstances which obviously give facility and advantage to the railway. One is, that the obstructions to the rolling motion of the wheels, produced by the inequalities of the surface, are very considerably less on a railway than on a road; less in the proportion of at least 1 to 20. This proportion, however, must depend much on the nature of the road with which the railway is compared. It is obvious that a well constructed road will offer less resistance than one ill constructed; and it is ascertained that the resistance of a Macadamised road is considerably more than that of a road well paved with stones· the decision of this question, therefore, must involve the consideration of another, viz. whether roads may not be constructed, by pavement or otherwise, smoother and better adapted to carriages moved in the manner of steam carriages than the roads now used for horse-power?

But besides the greater smoothness of railroads compared with turnpike roads, they have another advantage, which we suspect to have been considerably exaggerated by those who have opposed the project for steam carriages on turnpike roads. One of the laws of adhesion, long since developed by experiment, and known to scientific men, is, that it is greater between the surfaces of bodies of the same nature than between those of a different nature. Thus between two metals of the same kind it is greater than between two metals of different kinds. Between two metals of any kind it is greater than between metal and stone, or between metal and wood. Hence, the wheels of steam carriages running on a railroad have a greater adhesion with the road, and therefore offer a greater resistance to slip round without the advance of the carriage, than wheels would offer on a turnpike road; for on a railroad the iron tire of the wheel rests in contact with the iron rail, while on a common road the iron tire rests in contact with the surface of stone, or whatever material the road may be composed of. Besides this, the dust and loose matter which necessarily collect on a common road, when pressed between the wheels and the solid base of the road act somewhat in the manner of rollers, and give the wheels a greater facility to slip than if the road were swept clean, and the wheels rested in immediate contact with its hard surface. The truth of this observation is illustrated on the railroads themselves, where the adhesion is found to be diminished whenever the rails are covered with any extraneous matter, such as dust or moist clay. Although the adhesion of the wheels of a carriage with a common road, however, be less than those of the wheels of a steam carriage with a railroad, yet still the actual adhesion on turnpike roads is greater in amount than has been generally supposed, and is quite suffi cient to propel carriages dragging after them loads of large amount.

The relative facility with which carriages are propelled on railroads and turnpike roads equally affects any moving power, whether that of horses or steam engines ; and whether loads be propelled by the one power or the other, the railroad, as compared with the turnpike road, will always possess the same proportionate advantage ; and a given amount of power, whether of the one kind or the other, will always perform a quantity of work less in the same proportion on a turnpike road than on a railroad. But, on the other hand, the expense of original construction, and of maintaining the repairs of a railroad, is to be placed against the certain facility which it offers to draught.

In the attempts which have been made to adapt locomotive engines to turnpike roads, the projectors have aimed at the accomplishment of two objects : first, the construction of lighter and smaller engines ; and, secondly, increased power. These ends, it is plain, can only be attained, with our present knowledge, by the production of steam of very high temperature and pressure, so that the smallest volume of steam shall produce the greatest possible mechanical effect. The methods of propelling the carriage have been in general similar to that used in the railroad engines, viz. either by cranks placed on the axles, the wheels being fixed upon the same axles, or by connecting the piston-rods with the spokes of the wheels, as in the engine represented in fig. 55. In some carriages, the boiler and moving power, and the body of the carriage which bears the passengers, are placed on the same wheels. In others, the engine is placed on a separate carriage, and draws after it the carriage which transports the passengers, as is always the case on railways.

The chief difference between the steam engines used on railways, and those adapted to propel carriages on turnpike roads, is in the structure of the boiler. In the latter it is essential that, while the power remains undiminished, the boiler should be lighter and smaller. The accomplishment of this has been attempted by various contrivances for so distributing the water, as to expose a considerable quantity of surface in contact with it to the action of the fire ; spreading it in thin layers on flat plates ; inserting it between plates of iron placed at a small distance asunder, the fire being admitted between the intermediate plates ; dividing it into small tubes, round which the fire has play ; introducing it between the surfaces of cylinders placed one within another, the fire being admitted between the alternate cylinders,—have all been resorted to by different projecters.

(105.) First and most prominent in the history of the application of steam to the propelling of carriages on turnpike roads, stands the name of Mr. Goldsworthy Gurney, a medical gentleman and scientific chymist, of Cornwall. In 1822, Mr. *Gurney* succeeded Dr. Thompson as lecturer on chymistry at the Surrey Institution ; and, in consequence of the results of some experiments on heat, his attention was directed to the project of working steam carriages on common roads ; and since 1825 he has devoted his exertions in perfecting a steam engine capable of attaining the end he had in view. Numerous other projectors, as might have been expected, have followed in his wake. Whether they, or any of them, by better fortune, greater public support, or more powerful genius, may outstrip him in the career on which he has ventured, it would not, perhaps, at present, be easy to predict. But whatever be the event, to Mr. *Gurney* is due, and will be paid, the honour of first proving the practicability of the project; and in the history of the adaptation of the locomotive engine to common roads, his name will stand before all others in point of time, and the success of his attempts will be recorded as the origin and cause of the success of others in the same race.

The incredulity, opposition, and even ridicule, with which the project of Mr *Gurney* was met, are very remarkable. His views were from the first opposed by engineers, without one exception. The contracted habit of mind, sometimes produced by an education chiefly, if not exclusively, directed to a merely practical object, subsequently confirmed by exclusively practical pursuits, may, perhaps, in some degree, account for this. But, I confess, it has not been without surprise that I have observed, during the last ten years, the utter incredulity which has prevailed among men of general science on this subject,—an incredulity which the most unequivocal practical proof has scarcely yet dispelled. "Among scientific men," says Mr. *Gurney*, "my opinion had not a single supporter, with the exception of the late Dr. *Wollaston.*"

The mistake which so long prevailed in the application of locomotives on railroads, and which, as we have shown, materially retarded the progress of that invention, was shared by Mr. *Gurney*. Without reducing the question to the test of experiment, he took for granted, in his first attempts, that the adhesion of the wheels with the road was too slight to propel the carriage. He was assured, he says, by eminent engineers, that this was a point settled by actual experiment. It is strange, however, that a person of his quickness and sagacity did not inquire after the particulars of these "actual experiments." So, however, it was; and, taking for granted the inability of the wheels to propel, he wasted much labour and skill in the contrivance of levers and propellers, which acted on the ground in a manner somewhat resembling the feet of horses, to drive the carriage forward. After various fruitless attempts of this kind, the experience acquired in the trials to which they gave rise at last forced the truth upon his notice, and he found that the adhesion of the wheels was not only sufficient to propel the carriage heavily laden on level roads, but was capable of causing it to ascend all the hills which occur on ordinary turnpike roads. In this manner it ascended all the hills between London and Barnet, London and Stanmore,

Stanmore Hill, Brockley Hill, and mounted Old Highgate Hill, the last at one point rising one foot in nine.

It would be foreign to my present object to detail minutely all the steps by which Mr. *Gurney* gradually improved his contrivance. This, like other inventions, has advanced by a series of partial failures ; but it has at length attained that state, in which, by practice alone, on a more extensive scale, a further degree of perfection can be obtained.

(106.) The boiler of this engine is so constructed that there is no part of it, not even excepting the grate bars, in which metal exposed to the action of the fire is out of contact with water. If it be considered how rapidly the action of an intense furnace destroys metal when water is not present to prevent the heat from accumulating, the advantage of this circumstance will be appreciated. I have seen the bars of a new grate, never before used, melted in a single trip between Liverpool and Manchester ; and the inventor of another form of locomotive engine has admitted to me that his grate bars, though of a considerable thickness, would not last more than a week. In the boiler of Mr. *Gurney*, the grate bars themselves are tubes filled with water, and form, in fact, a part of the boiler itself. This boiler consists of three strong metal cylinders placed in a horizontal position one above the other. A section, made by a perpendicular or vertical plane, is represented in fig. 62. The ends of the three cylinders, just mentioned, are represented at D, H, and I. In the side of the lowest cylinder D are inserted a row of tubes, a ground plan of which is represented in fig. 63. These tubes, proceeding from the side of the lowest cylinder D, are inclined slightly upward, for a reason which I shall presently explain. From the nature of the section, only one of these tubes is visible in fig. 62, at C. The other extremities of these tubes at A are connected with the same number of upright tubes, one of which is expressed at E. The upper extremities G of these upright tubes are connected with another set of tubes K, equal in number, proceeding from G, inclining slightly upward, and terminating in the second cylinder H.

An end view of the boiler is exhibited in fig. 64, where the three cylinders are expressed by the same letters. Be

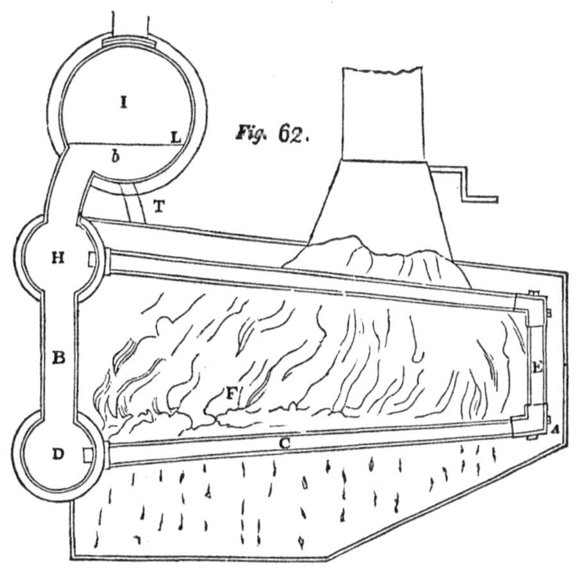

Fig. 62.

Fig. 63.

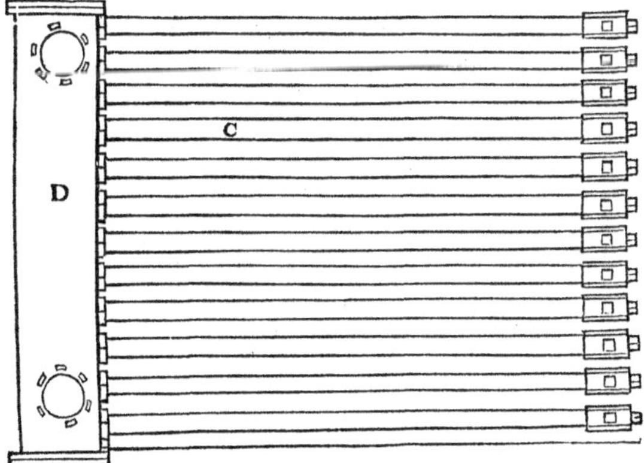

tween the cylinders D and H there are two tubes of communication B, and two similar tubes between the cylinders H and I. From the nature of the section these appear only as a single tube in fig. 62. From the top of the cylinder I proceeds a tube N, by which steam is conducted to the engine.

*Fig.* 64.

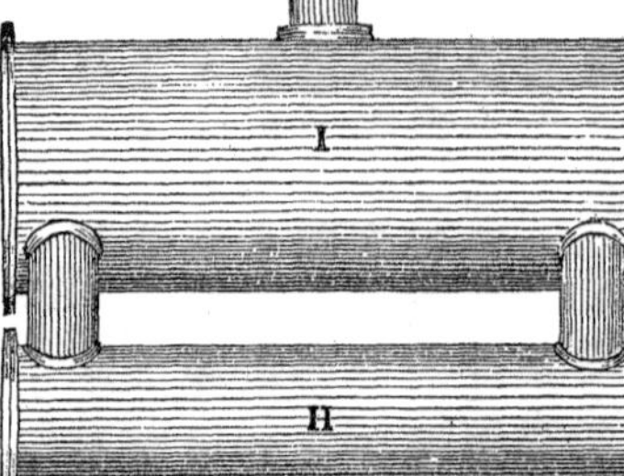


It will be perceived that the space F is enclosed on every side by a grating of tubes, which have free communication with the cylinders D and H, which cylinders have also a free communication with each other by the tubes B. It follows, therefore, that if water be supplied to the cylinder I, it will descend through the tubes, and first filling the cylinder D and the tubes C, will gradually rise in the tubes B and E, will next fill the tubes K and the cylinder H. The grating of water pipes C E K forms the furnace, the pipe C being the fire bars, and the pipes E and R be.ng the back and roof of the stove. The fire door, for the supply of fuel, appears at M, fig. 64 The flue issuing between the tubes F is conducted over the tubes K, and the flame and hot air are carried off through a chimney. That portion of the heat of the burning fuel, which in other furnaces destroys the bars of the grate, is here expended in heating the water contained in the tubes C. The radiant heat of the fire acts upon the tubes K, forming the roof of the furnace, on the tube E at the back of it, and partially on the cylinders D and H, and the tubes B. The draft of hot air and flame passing into the flue at A, acts upon the posterior surfaces of the tubes E, and the upper sides of the tubes K, and finally passes into the chimney.

As the water in the tubes C E K is heated, it becomes specifically lighter than water of a less temperature, and consequently acquires a tendency to ascend. It passes, therefore, rapidly into H. Meanwhile the colder portions descend, and the inclined positions of the tubes C and K give play to this tendency of the heated water, so that a prodigiously rapid circulation is produced, when the fire begins to act upon the tubes. When the water acquires such a temperature that steam is rapidly produced, steam bubbles are constantly formed in the tubes surrounding the fire; and if these remained stationary in the tubes, the action of the fire would not only decompose the steam, but render the tubes red-hot, the water not passing through them to carry off the heat. But the inclined position of the tubes, already noticed, effectually prevents this injurious consequence. A steam bubble which is formed either in the tubes C or K, having a tendency to ascend proportional to its lightness as compared with water, necessarily rushes upward; if in C toward A, and if in K toward H. But this motion of the steam is also aided by the rapid circulation of the water which is continually maintained in the tubes, as already explained, otherwise it might be possible, notwithstanding the levity of steam compared with water, that a bubble might remain in a narrow tube without rising. I notice this more particularly, because the burning of the tubes is a defect which has been erroneously, in my opinion, attributed to this boiler. To bring the matter to the test of experiment, I have connected two cylinders, such as D and H, by a system of glass tubes, such as represented at C E K. The rapid and constant circulation of the water was then made evident: bubbles of steam were formed in the tubes, it is true; but they passed with great rapidity into the upper cylinder, and rose to the surface, so that the glass tubes never acquired a higher temperature than that of the water which passed through them.

This I conceive to be the chief excellence of Mr. *Gurney's* boiler. It is impossible that any part of the metal of which it is formed can receive a greater temperature than that of the water which it contains; and that temperature, as is obvious, can be regulated with the most perfect certainty and precision. I have seen the tubes of this boiler, while exposed to the action of the furnace, after that action has continued for a long period of time, and I have never observed the soot which covers them to redden, as it would do if the tube attained a certain temperature.

Every part of the boiler being cylindrical, it has the form which, mechanically considered, is most favourable to strength, and which, within given dimensions, contains the greatest quantity of water. It is also free from the defects arising from unequal expansion, which are found to be most injurious in tubular boilers. The tubes C and K can freely expand in the direction of their length, without being loosened at their joints, and without straining any part of the apparatus; the tubes E, being short, are subject to a very slight degree of expansion; and it is obvious that the long tubes, with which they are connected, will yield to this without suffering a strain, and without causing any part of the apparatus to be loosened.

When water is converted into steam, any foreign matter which may be combined with it is disengaged, and is deposited on the bottom of the vessel in which the water is evaporated. All boilers, therefore, require occasional cleans ing, to prevent the crust thus formed from accumulating; and this operation, for obvious reasons, is attended with peculiar difficulty in tubular boilers. In the case before us, the crust of deposited matter would gather and thicken in the tubes C and K, and if not removed, would at length choke them. But besides this, it would be attended with a still worse effect; for, being a bad conductor, it would intercept the heat in its transit from the fire to the water, and would cause the metal of the tube to become unduly heated. Mr. *Gurney* of course foresaw this inconvenience, and contrived an ingenious chymical method of removing it by occasionally injecting through the tubes such an acid as would combine with the deposite, and carry it away. This method was perfectly effectual; and although its practical application was found to be attended with difficulty in the hands of common workmen, Mr. *Gurney* was persuaded to adhere to it by the late Dr. *Wollaston*, until experience proved the impossibility of getting it effectually performed, under the circumstances in which boilers are commonly used. Mr. *Gurney* then adopted the more simple, but not less effectual, method of removing the deposite by mechanical means. Opposite the mouths of the tubes, and on the other side of the cylinders D and H, are placed a number of holes, which, when the boiler is in use, are stopped by pieces of metal screwed into them. When the tubes require to be cleaned these stoppers are removed, and an iron scraper is introduced through the holes into the tubes, which, being passed backward and forward, removes the deposite. The boiler may be thus cleaned by a common labourer in half a day, at an expense of about 1*s.* 6*d.*

The frequency of the periods at which a boiler of this kind requires cleaning must depend, in a great degree, on the nature of the water which is used; one in daily use with the water of the river Thames would not require cleaning more than once in a month. Mr. *Gurney* states that with water of the most unfavourable description, once a fortnight would be sufficient.

(107.) In the more recent boilers constructed by Mr. *Gurney*, he has maintained the draught through the furnace, by the method of projecting the waste steam into the chimneys; a method so perfectly effectual, that it is unlikely to be superseded by any other. The objection which has been urged against it in locomotive engines, working on turnpike roads, is, that the noise which it produces has a tendency to frighten horses.

In the engines on the Liverpool road, the steam is allowed to pass directly from the eduction pipe of the cylinder to the chimney, and it there escapes in puffs corresponding with the alternate motion of the pistons, and produces a noise, which, although attended with no inconvenience on the railroad, would certainly be objectionable on turnpike roads. In the engine used in Mr. *Gurney's* steam carriage, the steam which passes from the cylinders is conducted to a receptacle, which he calls a blowing box. This box serves the same purpose as the upper chamber of a smith's bellows. It receives the steam from the cylinders in alternate puffs, but lets it escape into the chimney in a continued stream by a number of small jets. Regular draught is by this means produced, and no noise is perceived. Another exit for the steam is also provided, by which the conductor is enabled to increase or diminish, or to suspend altogether, the draught in the chimney, so as to adapt the intensity of the fire to the exigencies of the road. This is a great convenience in practice; because, on some roads, a draught is scarcely required, while on others a powerful blast is indispensable.

Connected with this blowing box, is another ingenious apparatus of considerable practical importance. The pipe through which the water which feeds the boiler is conducted to it from the tank is carried through this blowing box, within which it is coiled in a spiral form, so that an extensive thread of the feeding water is exposed to the heat of the waste steam which has escaped from the cylinders, and which is enclosed in this blowing box. In passing through this pipe the feeding water is raised from the ordinary temperature of about 60° to the temperature of 212°. The fuel necessary to accomplish this is, therefore, saved ; and the amount of this is calculated at 1-6th of all that is necessary to evaporate the water. Thus, 1-6th of the expense of fuel is saved. But, what is much more important in a locomotive engine, a portion of the weight of the engine is saved without any sacrifice of its power. There is still another great advantage attending this process. The feeding water in the worm just mentioned, while it takes up the heat from the surrounding steam in the blowing box, condenses 1-6th of the waste steam, which is thence conducted to the tank, from which the feeding water is pumped, saving in this manner 1-6th in weight and room of the water necessary to be carried in the carriage for feeding the boiler.*

So far as the removal of all inconvenience arising from noise, this contrivance has been proved by experience to be perfectly effectual.†

In all boilers, the process of violent ebullition causes a state of agitation in the water, and a number of counter currents, by which, as the steam is disengaged from the surface of the water, it takes with it a considerable quantity of water in mechanical mixture. If this be carried through the cylinders, since it possesses none of the qualities of steam, and adds nothing to the power of the vapour with which it is combined, it causes an extensive waste of heat and water,

* In boilers constructed for stationary purposes, or for steam navigation, the steam pipe, after it has passed through the blowing box, is continued and made to form a series of returned flues over the boiler, so as to take up the waste heat after it has passed the boiler, and before it reaches the chimney. But in locomotive engines for common roads, it has been found by experience, that the power gained by the waste heat is not sufficient to propel the weight of the material necessary for taking it up.

† See Report of the Commons.

and produces other injurious effects. In every boiler, therefore, some means should be provided for the separation of the water thus suspended in the steam, before the steam is conducted to the cylinder. In ordinary plate boilers, the large space which remains above the surface of the water serves this purpose. The steam being there subject to no agitation or disturbance, the water mechanically suspended in it descends by its own gravity, and leaves pure steam in the upper part. In the small tubular boilers, this has been a matter, however, of greater difficulty. The contracted spaces in which the ebullition takes place causes the water to be mixed with the steam in a greater quantity than could happen in common plate boilers: and the want of the same steam-room renders the separation of the water from the steam a matter of some difficulty. These inconveniences have been overcome by a succession of contrivances of great ingenuity. I have already described the rapid and regular circulation effected by the arrangement of the tubes. By this a regularity in the currents is established, which alone has a tendency to diminish the mixture of water with the steam. But in addition to this, a most effectual method of separation is provided in the vessel I, which is a strong iron cylinder of some magnitude, placed out of the immediate influence of the fire. A partial separation of the steam from the water takes place in the cylinder H; and the steam with the water mechanically suspended in it, technically called moist steam, rises into the *separator* I. Here, being free from all agitation and currents, and being, in fact, quiescent, the particles of water fall to the bottom, while the pure steam remains at the top. This separator, therefore, serves all the purposes of the steam-room above the surface of the water in the large plate boilers. The dry steam is thus collected and ready for the supply of the engine through the tube N, while the water, which is disengaged from it, is collected at the bottom of the separator, and is conducted through the tube T to the lowest vessel D, to be again circu lated through the boiler.

The pistons of the engine work on the axles of the hind wheels of the carriage which bears the engine, by cranks, as in the locomotives on the Manchester railway, so that the axle is kept in a constant state of rotation while the engine is at work. The wheels placed on this axle are not permanently fixed or keyed upon it, as in the Manchester locomotives; but they are capable of turning upon it in the same manner as ordinary carriage wheels. Immediately within these wheels there are fixed upon the axles two projecting spokes or levers, which revolve with the axle, and which take the position of two opposite spokes of the wheel. These may be occasionally attached to the wheel or detached from it; so that they are capable of compelling the wheels to turn with the axle, or leaving the axle free to turn independent of the wheel, or the wheel independent of the axle, at the pleasure of the conductor. It is by these levers that the engine is made to propel either or both of the wheels. If both pairs of spokes are thrown into connexion with the wheels, the crank shaft or axle will cause both wheels to turn with it, and in that case the operation of the carriage is precisely the same as those of the locomotives already described upon the Liverpool and Manchester line; but this is rarely found to be necessary, since the adhesion of one wheel with the road is generally sufficient to propel the carriage, and consequently only one pair of these fixed levers are generally used, and the carriage propelled by only one of the two hind wheels. The fore wheels of the carriage turn upon a pivot similar to those of a four-wheeled coach. The position of these wheels is changed at pleasure by a simple pinion and circular rack, which is moved by the conductor, and in this manner the carriage is guided with precision and facility.

The force of traction necessary to propel a carriage upon common roads must vary with the variable quality of the road, and consequently the propelling power, or the pressure upon the pistons of the engine, must be susceptible of a corresponding variation; but a still greater variation becomes necessary from the undulations and hills which are upon all ordinary roads. This necessary change in the intensity of the impelling power is obtained by restraining the steam in the boiler by the throttle valve, as already described in the locomotive engines on the railroad. This principle, however, is carried much further in the present case. The steam in the boiler may be at a pressure of from 100 to 200 lbs. on the square inch; while the steam on the working piston may not exceed 30 or 40 lbs. on the inch. Thus an immense increase of power is always at the command of the conductor; so that when a hill is encountered, or a rough piece of road, he is enabled to lay on power sufficient to meet the exigency of the occasion.

The two difficulties which have been always apprehended in the practical working of steam carriages upon common roads are, first, the command of sufficient power for hills and rough pieces of road; and, secondly, the apprehended insufficiency of the adhesion of the wheels with the road to propel the carriage. The former of these difficulties has been met by allowing steam of a very great pressure to be constantly maintained in the boiler with perfect safety. As to the second, all experiments tend to show that there is no ground for the supposition that the adhesion of the wheels is in any case insufficient for the purposes of propulsion. Mr. *Gurney* states, that he has succeeded in driving carriages thus propelled up considerable hills on the turnpike roads about London. He made a journey to Barnet with only one wheel attached to the axle, which was found sufficient to propel the carriage up all the hills upon that road. The same carriage, with only one propelling wheel, also went to Bath, and surmounted all the hills between Cranford Bridge and Bath, going and returning.

A double stroke of the piston produces one revolution of

the propelling wheels, and causes the carriage to move through a space equal to the circumference of those wheels. It will therefore be obvious that the greater the diameter of the wheels, the better adapted the carriage is for speed; and, on the other hand, wheels of smaller diameter are better adapted for power. In fact, the propelling power of an engine on the wheels will be in the inverse proportion to their diameter. In carriages designed to carry great weights at a moderate speed, smaller wheels will be used; while in those intended for the transport of passengers at considerable velocities, wheels of at least five feet in diameter are most advantageous.

Among the numerous popular prejudices to which this new invention has given rise, one of the most mischievous in its effects, and most glaring in its falsehood, is the notion that carriages thus propelled are more injurious to roads than carriages drawn by horses. This error has been most clearly and successfully exposed in the evidence taken before the committee of the House of Commons upon steam carriages. It is there fully demonstrated, not only that carriages thus propelled did not wear a turnpike road more rapidly than those drawn by horses, but that, on the other hand, the wear by the feet of horses is far more rapid and destructive than any which could be produced by the wheels of carriages. Steam carriages admit of having the tires of the wheels broad, so as to act upon the road more in the manner of rollers, and thereby to give consistency and firmness to the material of which the road is composed. The driving wheels, being fully proved not to slip upon the road, do not produce any effects more injurious than the ordinary rolling wheels; consequently the wear occasioned by a steam carriage upon a road is not more than that produced by a carriage drawn by horses of an equivalent weight and the same or equal tires; but the wear produced by the pounding and digging of horses' feet in draught is many times greater than that produced by the wear of any carriage. Those who still have doubts upon this subject, if there be any such persons, will be fully satisfied by referring to the evidence which accompanies the report of the committee of the House of Commons, printed in October, 1831. In that report they will not only find demonstrative evidence that the introduction of steam carriages will materially contribute to the saving in the wear of turnpike roads, but also that the practicability of working such carriages with a great saving to the public—with great increase of speed and other conveniences to the traveller—is fully established.

The weight of machinery necessary for steam carriages is sometimes urged as an objection to their practical utility. Mr. *Gurney* states, that, by successive improvements in the details of the machinery, the weight of his carriages, without losing any of the propelling power, may be reduced to 35 cwt., exclusive of the load and fuel and water: but thinks that it is possible to reduce the weight still further.

A steam carriage constructed by Mr. *Gurney*, weighing 35 cwt., working for 8 hours, is found, according to his statement, to do the work of about 30 horses. He calculates that the weight of his propelling carriage, which would be capable of drawing 18 persons, would be equal to the weight of 4 horses; and the carriage in which these persons would be drawn would have the same weight as a common stage coach capable of carrying the same number of persons. Thus the weight of the whole—the propelling carriage and the carriage for passengers taken together—would be the same with the weight of a common stage coach, with 4 horses inclusive.

(108.) There are two methods of applying locomotives upon common roads to the transport of passengers or goods; the one is by causing the locomotive to carry, and the other to draw the load; and different projectors have adopted the one and the other method. Each is attended with its advantages and disadvantages. If the same carriage transport the engine and the load, the weight of the whole will be less in proportion to the load carried; also a greater pressure may be produced on the wheels by which the load is propelled. It is also thought that a greater facility in turning and guiding the vehicle, greater safety in descending the hills, and a saving in the original cost, will be obtained. On the other hand, when the passengers are placed in the same carriage with the engine, they are necessarily more exposed to the noise of the machinery and to the heat of the boiler and furnace. The danger of explosion is so slight, that, perhaps, it scarcely deserves to be mentioned; but still *the apprehension* of danger on the part of the passengers, even though grondless, should not be disregarded. This apprehension will be obviously removed or diminished by transferring the passengers into a carriage separate from the engine; but the greatest advantage of keeping the engine separate from the passengers is the facility which it affords of changing one engine for another in case of accident or derangement on the road, in the same manner as horses are changed at the different stages: or, if such an accident occur in a place where a new engine cannot be procured, the load of passengers may be carried forward by horse, until it is brought to some station where a locomotive may be obtained. There is also an advantage arising from the circumstance that when the engines are under repair, or in process of cleaning, the carriages for passengers are not necessarily idle. Thus the same number of carriages for passengers will not be required when the engine is used to draw as when it is used to carry.

In case of a very powerful engine being used to carry great loads, it would be quite impracticable to place the engine and loads on four wheels, the pressure being such as no turnpike road could bear. In this case it would be indispensably necessary to place a part of the load at least upon separate carriages to be drawn by the engine.

In the comparison of carriages propelled by steam with carriages drawn by horses, there is no respect in which the advantage of the former is so apparent as the safety afforded to the passenger. Steam power is under the most perfect control, and a carriage thus propelled is capable of being guided with the most admirable precision. It is also capable of being stopped almost suddenly, whatever be its speed; it is capable of being turned within a space considerably less than that which would be necessary for four-horse coaches. In turning sharp corners, there is no danger, with the most ordinary care on the part of the conductor. On the other hand, horse-power, as is well known, is under very imperfect control, especially when horses are used, adapted to that speed which at present is generally considered necessary for the purpose of travelling. "The danger of being run away with and overturned," says Mr. *Farey*, in his evidence before the House of Commons, "is greatly diminished in a steam coach. It is very difficult to control four such horses as can draw a heavy stage coach ten miles an hour, in case they are frightened or choose to run away; and, for such quick travelling, they must be kept in that state of courage that they are always inclined to run away, particularly down hill, and at sharp turns in the road. Steam-power has very little corresponding danger, being perfectly controllable, and capable of having its power reversed to retard in going down hill. It must be carelessness that would occasion the overturning of a steam carriage. The chance of breaking down has been hitherto considerable, but it will not be more than in stage coaches when the work is truly proportioned and properly executed. The risk from explosion of the boiler is the only new cause of danger, and that I consider not equivalent to the danger from horses."

That the risk of accident from explosion is very slight indeed, if any such risk exists, may be proved from the fact that the boilers used on the Liverpool and Manchester railroad being much larger, and, in proportion, inferior in strength to those of Mr. *Gurney*, and other steam carriage projectors, have never yet been productive of any injurious consequences by explosion, although they have frequently burst. I have stood close to a locomotive on the railroad when the boiler burst. The effect was that the water passed through the tubes into the fire and extinguished it, but no other consequence ensued.

In fig. 65 is represented the appearance of a locomotive of Mr. *Gurney's*, drawing after it a carriage for passengers.

*Fig.* 65.

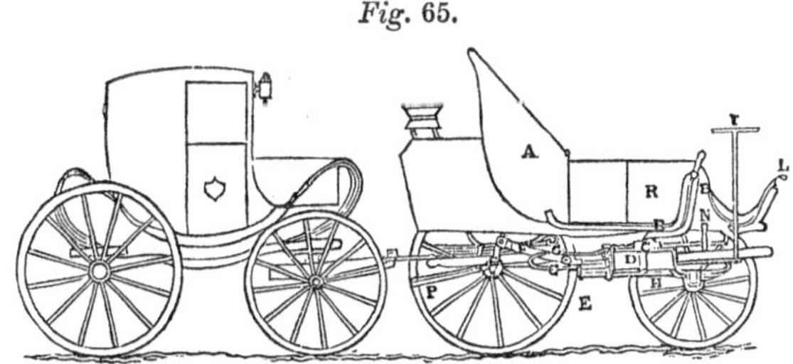


(109.) One of the greatest difficulties which locomotives upon a turnpike road have to encounter is the ascent of very steep hills, for it is agreed upon all hands that hills of very moderate inclinations present no difficulty which may not be easily overcome, even in the present state of our knowledge. The fact of Mr. *Gurney* having propelled his carriage up Old Highgate Hill, when the apparatus was in a much more imperfect state than that to which it has now attained, establishes the mere question of the possibility of overcoming the difficulty; but it remains still to be decided whether the inconvenience caused by providing means of meeting the exigency of very steep hills may not be greater than the advantage of being able to surmount them can compensate for: and Mr. *Farey*, whose authority upon subjects of this kind is entitled to the highest respect, thinks that it is upon the whole more advantageous to provide, at very steep hills, post horses to assist the steam carriage up them, than to incur the inconvenience of providing the necessary power and strength of machinery for occasions which at best but rarely occur. If the question merely referred to the command of motive power, it appears to me that Mr. *Gurney's* boiler would be amply sufficient to supply all that could be required for any hills which occur upon turnpike roads; but it is not to be forgotten, that not merely an ample supply of motive power, but also a strength and weight in the machinery proportionate to the power to be exerted, is indispensably necessary. The strength and weight necessary to ascend a very steep hill will be considerably greater than that which is necessary for a level road, or for hills of moderate inclinations; and it follows that if we ascend those steep hills by the unaided power of the locomotive, we must load the engine with all the weight of machinery requisite for such emergencies, such additional weight being altogether unnecessary, and therefore a serious impediment, upon all other parts of the road, inasmuch as it must exclude an equivalent weight of goods or passengers, which might otherwise be transported, and thereby in fact diminish proportionally the efficiency of the machine. It is right, however, to observe, that this is a point upon which a difference of opinion is entertained by persons equally competent to form a judgment, and that some consider that it is practicable to construct an engine without inconvenient weight which will ascend all the hills which occur upon turnpike roads.

However this may be, the difficulty is one which the improved system of roads in England renders of a comparatively trifling nature. If horses were resorted to as the means of assistance up such hills as the engine would be incapable of surmounting, such aid would not be requisite more than twice or thrice upon the mail-coach road between London and Holyhead; and the same may be said of the roads connecting the greatest points of intercourse in the kingdom. Such hills as the ascent at Pentonville upon the New Road, the ascent in St. James's Street, the ascent from Waterloo Place to the County Fire Office, the ascent at Highgate Archway, present no difficulty whatever. It is only Old Highgate Hill, and hills of a similar kind, which would ever require a supplv of horses in aid of the engine.

I therefore incline to agree with Mr. *Farey*, that, at least for the present, it will be more expedient to construct carriages adapted to surmount moderate hills only, and to provide post horses in aid of the extreme emergencies to which I have just alluded.

(110.) In the boiler to be used in the steam carriage projected by Mr. *Walter Hancock*, the subdivision of the water is accomplished by dividing a case or box by a number of thin plates of metal like a galvanic battery, the water being allowed to flow between every alternate pair of plates at E, fig. 66, and the intermediate spaces H forming the flue through which the flame and hot air are propelled.

In fact, a number of thin plates of water are exposed on both sides to the most intense action of flame and heated air; so that steam of a high pressure is produced in great abundance and with considerable rapidity. The plates forming the boiler are bolted together by strong iron ties, extending across the boiler, at right angles to the plates, as represented

*Fig.* 66.

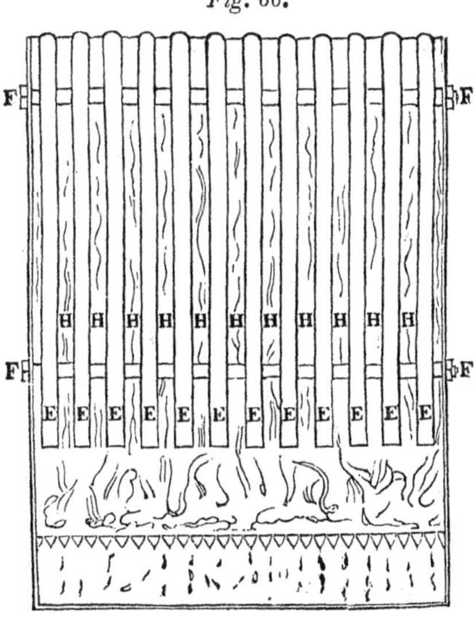

in the figure. The distance between the plates is two inches.

There are ten flat chambers of this kind for water, and intermediately between them ten flues. Under the flues is the fireplace, or grate ; containing six square feet of fuel in vivid combustion. The chambers are all filled to about two-thirds of their depth with water, and the other third is left for steam. The water-chambers, throughout the whole series, communicate with each other both at top and bottom, and are held together by two large bolts. By releasing these bolts, at any time, the chambers fall asunder; and by screwing them up, they may be all made tight again. The water is supplied to the boiler by a forcing-pump, and the steam issues from the centre of one of the flues at the top.

These boilers are constructed to bear a pressure of 400 or 500 lbs. on the square inch ; but the average pressure of the steam on the safety valve is from 60 to 100. There are 100 square feet of surface in contact with the water exposed to the fire. The stages which such an engine performs are eight miles, at the end of which a fresh supply of fuel and water are taken in. It requires about two bushels of coke for each stage.

The steam carriage of Mr. *Hancock* differs from that of Mr. *Gurney* in this,—that in the former the passengers and engine are all placed on the same carriage. The boiler is placed behind the carriage ; and there is an engine-house between the boiler and the passengers, who are placed on the fore part of the vehicle ; so that all the machinery is behind them. The carriages are adapted to carry 14 passengers, and weigh, exclusive of their load, about $3\frac{1}{2}$ tons, the tires of the wheels being about $3\frac{1}{2}$ inches in breadth. Mr. *Hancock* states, that the construction of his boiler is of such a nature, that, even in the case of bursting, no danger is to be apprehended, nor any other inconvenience than the stoppage of the carriage. He states that, while travelling about 9 miles an hour, and working with a pressure of about

100 lbs. on the square inch, loaded with 13 passengers, the carriage was suddenly stopped. At first the cause of the accident was not apparent; but, on opening one of the cocks of the boiler, it was found that it contained neither steam nor water. Further examination proved that the boiler had burst. On unscrewing the bolts, it was found there were several large holes in the plates of the water-chamber, through which the water had flowed on the fire; but neither noise nor explosion, nor any dangerous consequences ensued.

This boiler has some obvious defects. It is evident that thin flat plates are the form which, mechanically considered, is least favourable to strength; nor does it appear that any material advantage is gained to compensate for this by the magnitude of the surface exposed to the action of the fire. It is a great defect that a part of the surface of each of the plates is exposed to the action of the fire while it is out of contact with the water; in fact, in the upper part of the spaces marked E, fig. 66, steam only is contained. It has been observed by engineers, and usually shown by experiment, that if steam be heated on the surface of water it will be decomposed, and its elasticity destroyed; this is not the only evil connected with the arrangement, for on this part of the metal, nevertheless, the fire acts,—with less intensity, it is true, than on that part which contains the water,—but still with sufficient intensity to destroy the metal. Mr. *Hancock* appears to have attempted to remedy this defect by occasionally inverting the position of the flat chambers, placing that which at one time was at the bottom at the top, and *vice versâ*. This may equalize the wear produced by the action of the fire upon the metal out of contact with water, but still the wear on the whole will not be less rapid. There appears to be no provision or space for separating the steam from the water with which it is charged; in fact, there are no means in this engine of discharging the function of Mr. *Gurney's* separator. This will be found to produce considerable waste and loss of power in practice.

The bars upon which the fire rests are of solid metal; and such is the intense heat to which they are subject, that, in an engine constantly at work, it is unlikely that they will last, without being renewed, more than about a week, if so much. The draft is maintained in this engine by means of a revolving fan worked by the engine. This, perhaps, is one of the greatest defects as compared with other locomotives. The quantity of power requisite to work this bellows, and of which the engine is robbed, is very great. This defect is so fatal, that I consider it is quite impossible that the ingenious inventor can persevere in the use of it. Mr. *Hancock* has abandoned the use of the cranks upon his working axle, and has substituted an endless chain and rag-wheels. This also appears to me defective, and a source by which considerable power is lost. On the other hand, however, the weakness of the axle which is always produced by cranks is avoided.

(111.) Mr. *Nathaniel Ogle* of Southampton obtained a patent for a locomotive carriage, and worked it for some time experimentally ; but as his operations do not appear to have been continued, I suppose he was unsuccessful in fulfilling those conditions, without which the machine could not be worked with economy and profit. In his evidence before a committee of the House of Commons, he has thus described his contrivance :—

" The base of the boiler and the summit are composed of cross pieces, cylindrical within and square without ; there are holes bored through these cross pieces, and inserted through the whole is an air tube. The inner hole of the lower surface, and the under hole of the upper surface, are rather larger than the other ones. Round the air tube is placed a small cylinder, the collar of which fits round the larger aperture on the inner surface of the lower frame, and the under surface of the upper frame-work. These are both drawn together by screws from the top; these cross pieces are united by connecting pieces, the whole strongly bolted together; so that we obtain, in one-tenth of the space, and with one-tenth of the weight, the same heating surface and power as is now obtained in other and low-pressure boilers, with incalculably greater safety. Our present experimental boiler contains 250 superficial feet of heating surface in the space of 3 feet 8 inches high, 3 feet long, and 2 feet 4 inches broad, and weighs about 8 cwt. We supply the two cylinders with steam, communicating by their pistons with a crank axle, to the ends of which either one or both wheels are affixed, as may be required. One wheel is found to be sufficient, except under very difficult circumstances, and when the elevation is about one foot in six, to impel the vehicle forward.

"The cylinders of which the boiler is composed are so small as to bear a greater pressure than could be produced by the quantity of fire beneath the boiler; and if any one of these cylinders should be injured by violence, or any other way, it would become merely a safety valve to the rest. We never, with the greatest pressure, burst, rent, or injured our boiler; and it has not once required cleaning, after having been in use twelve months."

(112.) Dr. *Church* of Birmingham has obtained a succession of patents for contrivances connected with a locomotive engine for stone roads; and a company consisting of a considerable number of individuals, possessing sufficient capital, has been formed in Birmingham for carrying into effect his designs, and working carriages on his principle. The present boiler of Dr. *Church* is formed of copper. The water is contained between two sheets of copper, united together by copper nails, in a manner resembling the way in which the cloth forming the top of a mattrass or cushion is united with the cloth which forms the bottom of it, except that the nails or pins, which bind the sheets of copper, are much closer together. The water, in fact, seems to be "quilted" or "padded" in between two sheets of thin copper. This double sheet of copper is formed into an oblong rectangular box, the interior of which is the fireplace and ash-pit, and over the end of which is the steam-chest. The great extent of surface exposed to the immediate action of the fire causes steam to be produced with great rapidity.

An obvious defect which such a boiler presents is the difficulty of removing from it any deposite or incrustation, which may collect between the sheets of copper so closely and intricately connected. Dr. Church proposes to effect this, when it is required, by the use of an acid, which will combine readily with the incrustation, and by which the boiler may therefore be washed. This method of cleansing boilers was recommended by Dr. Wollaston to Mr. Gurney, who informed me, however, that he found that it was not practicable in the way in which boilers must commonly be used.

I apprehend, also, that the spaces between the sheets of copper, in Dr. Church's boiler just described, will hardly permit the steam bubbles which will be formed to escape with sufficient facility into the steam-chest; and being retained in that part of the boiler which is exposed to the action of the fire, the metal will be liable to receive an undue temperature.

I have, however, seen this engine working, and its performance was very satisfactory.

(113.) Various other projects for steam carriages on common roads are in various degrees of advancement, among which may be mentioned those of Messrs. Maudslay and Field, Col. Macerone, and Mr. Russell, of Edinburgh ; but our limits compel us to omit any detailed account of them.

# CHAPTER XII.

## STEAM NAVIGATION.

Propulsion by Paddle Wheels.—Manner of driving them.—Marine Engine.—Its Form and Arrangement.—Proportion of its Cylinder.—Injury to Boilers by Deposites and Incrustation.—Not effectually removed by *blowing out*.—Mr. Samuel Hall's Condenser.—Its Advantages.—Originally suggested by Watt.—Hall's *Steam Saver*.—Howard's Vapour Engine.—Morgan's Paddle Wheels.—Limits of steam Navigation.—Proportion of Tonnage to Power.—Average Speed.—Consumption of Fuel.—Iron Steamers.—American Steam Raft.—Steam Navigation to India.—By Egypt and the Red Sea to Bombay.—By same Route to Calcutta.—By Syria and the Euphrates to Bombay.—Steam Communication with the United States from the west Coast of Ireland to St. John's, Halifax, and New York.

(114.) Among the various ways in which the steam engine has ministered to the social progress of our race, none is more important and interesting than the aid it has afforded to navigation. Before it lent its giant powers to that art, locomotion over the waters of the deep was attended with a degree of danger and uncertainty, which seemed so necessary and so inevitable, that, as a common proverb, it became the type and representative of every thing else which was precarious and perilous. The application, however, of steam to navigation has rescued the mariner from much of the perils of the winds and waves; and even in its actual state, apart from the improvements which it is still likely to receive, it has rendered all voyages of moderate length as safe and as regular as journeys over land. We are even now upon the brink of such improvements as will probably so extend the powers of the steam engine as to render it available as the means of connecting the most distant points of the earth.

The manner in which the steam engine is commonly applied to propel vessels must be so familiar as to require but short explanation. A pair of wheels, like common under-shot water-wheels, bearing on their rims a number of flat boards called *paddle boards*, are placed one at each side of the vessel, in such a position that when the vessel is immersed to her ordinary depth the lowest paddle boards shall be submerged. These wheels are fixed upon a shaft, which is made to revolve by cranks placed upon it, in the same manner as the fly wheel of a common steam engine is turned. It is now the invariable custom to place in steam vessels two engines, each of which works a crank: these two cranks are placed at right angles to each other, in the same manner as the cranks already described upon the working axles of locomotive engines. When either crank is at its dead point, the other is in full activity, so that the necessity for a fly wheel is superseded. The engines may be either condensing or high-pressure engines; but in Europe the low-pressure condensing engine has been invariably used for nautical purposes. In the United States, where steam navigation had its origin, and where it was, until a recent period, much more extensively practised than in Europe, less objection was felt to the use of high-pressure engines; and their limited bulk, their small original cost, and simplicity of structure, strongly recommended them, more especially for the purposes of river navigation.

(*g*) The original type of nearly all the engines used in steam navigation was that constructed at Soho by Watt and Bolton for Mr. Fulton, and first used by him upon the Hudson river. This had the beam below the piston-rod, as in the English boat-engines, but the cylinder above deck, as in the American. From this primitive form, the two nations have diverged in opposite directions. The Americans, navigating rivers, and having speed for their principal object, have not hesitated to keep the cylinder above deck, and have lengthened the stroke of the piston in order to make the power cut on a more advantageous point of the wheel. Compactness has been gained by the suppression of the working beam.

On the other hand, the English, having the safe navigation f stormy seas as their more important object, have shortened the cylinder in order that the piston-rod may work wholly under the deck, and the arrangement of Fulton's working beam has been retained by them. In this way there can be no doubt, that they have lost the power of obtaining equal speed from a given expenditure of power, and those conversant in the practice and theory of stowing ships may well doubt whether security is not also sacrificed.—A. E.

The arrangement of the parts of the maritime engine differs, in some respects, from that of the land engine. Wan of room renders greater compactness necessary; and in order to diminish the height of the machine, the working beam is transferred from above the cylinder to below it. In fact, there are two beams, one at each side of the engine, which are connected by a parallel motion with the piston, the rods of the parallel motion extending from the lower part of the engine to the top of the piston-rod. The working end of the beam is connected with the crank by a connecting rod, presented upward instead of downward, as in the land engine. The proportion of the length and diameters of the cylinders differ from those of land engines for a like reason: to save height, short cylinders with large diameters are used. Thus, in an engine of 200 horse power, the length of the cylinder is sometimes 60 inches, and its diameter 53 inches: the valves and the gearing which work them, the air-pump, condenser, and other parts of the machine, do not differ materially from those already described in the land engines.

(*h*) The action of machinery may be rendered more equable by using two engines, each of half the power, instead of a single one. If one of these be working with its maximum force when the other is changing the direction of its motion, the result of their joint action will be a force nearly constant. Such a combination was invented by Mr. Francis Ogden, and has been used in several steamboats constructed under his directions. It would however be far more valuable in other cases, particularly where great uniformity in the velocity is indispensable.

This method has now become almost universal in the engines used in the English steamboats, each of which has usually two, both applied to the same shaft, and therefore capable of being used singly or together to turn the paddle wheels. In the American steamboats, although two engines have been often applied, each usually acts upon no more than one of the wheels. We can see no other good reason for this, than that our engineers do not wish to be thought to copy Mr. Ogden.—A. E.

The nature of the work which the marine engine has to perform is such, that great regularity of action is neither necessary nor possible. The agitation of the surface of the sea will cause the immersion of the paddle wheels to vary very much, and the resistance to the engine will undergo a corresponding change : the governor, and other parts of the apparatus already described, contrived for imparting to the engine that extreme regularity which is indispensable in its application to manufactures, are therefore here omitted ; and nothing is introduced except what is necessary to maintain the engine in its full working power.

It is evident that it must be a matter of considerable importance to reduce the space occupied by the machinery on board a vessel to the least possible dimensions. The marine boilers, therefore, are constructed so as to yield the necessary quantity of steam with the smallest practical dimensions. With this view a much more extensive surface in proportion to the size of the boiler is exposed to the action of the fire. In fact, the flues which carry off the heated air to the chimney are conducted through the boiler, so as to act upon the water on every side in thin oblong shells, which traverse the boiler backward and forward repeatedly, until finally they terminate in the chimney. By this arrangement the original expense of the boilers is very considerably increased; but, on the other hand, their steam-producing power is also greatly augmented; and from experiments lately made by Mr. Watt at Birmingham, it appears that they work with an economy of fuel compared with common land boilers in the proportion of about two to three. Thus they have the additional advantage of saving the tonnage as well as the expense of one-third of the fuel.

One of the most formidable difficulties which has been encountered in applying the steam engine to the purposes of navigation has arisen from the necessity of supplying the boiler with sea water, instead of pure fresh water. This water (also used for the purpose of condensation) being injected into the condenser, and mixed with the condensed steam, is conducted as feeding water into the boiler.

The salt contained in the sea water, not being evaporated, remains in the boiler. In fact, it is separated from the water in the same manner as by the process of distillation. As the evaporation in the boiler is continued, the proportion of salt contained in the water is, therefore, constantly increased, until a greater proportion is accumulated than the water is capable of holding in solution; a deposition of salt then commences, and is lodged in the cavities at the bottom of the boiler. The continuance of this process, it is evident, would at length fill the boiler with salt.

But besides this, under some circumstances, a deposition of lime* is made, and a hard incrustation is formed on the inner surface of the boiler. In some situations, also, sand and mud are received into the boiler, being suspended in the water pumped in for feeding it. All these substances,

* Ten thousand grains of pure sea water contain muriate of soda 220 grs., sulphate of soda 33 grs., muriate of magnesia 42 grs., and muriate of lime 8 grs.

whether deposited in a loose form in the lower parts of the boiler or collected in a crust on its inner surface, form obstructions to the passage of heat from the fire to the water. The crust thus formed is not unfrequently an inch or more in thickness, and so hard that good chisels are broken in removing it. The heat more or less intercepted by these substances collects in the metal of the boiler, and raises it to a temperature far exceeding that of the water within. It may even, if the incrustation be great, be sufficient to render the boiler red-hot. These circumstances occasion the rapid wear of the boiler, and endanger its safety by softening it.

The remedy which has generally been adopted to remove or diminish these injurious effects consists in allowing a stream of hot water continually to flow from the boiler, and supplying from the feed pipe a corresponding portion of cold water. While the hot water which flows from the boiler in this case contains, besides its just proportion of salt, that portion which has been liberated from the water converted into vapour, the cold water which is supplied through the feed pipe contains less than its just proportion of salt, since it is composed of the natural sea water, mixed with the condensed steam, which latter contains no salt. In this manner, the proportion of the salt in the boiler may be prevented from accumulating; but this is attended with considerable inconvenience and loss. It is evident that the discharge of the hot water, and the introduction of so considerable a quantity of cold water, entails upon the machine a great waste of fuel, and, consequently, renders it necessary that the vessel should be supplied with a much larger quantity of coals than are merely necessary for propelling it. In long voyages, where this inconvenience is most felt, this is a circumstance of obvious importance. But besides the waste of fuel, the speed of the vessel is diminished by the rate of evaporation in the boiler being checked by the constant stream of cold water flowing into it. This process of discharging the water, which is called *blowing out*, is only practised occasionally. In the Admiralty steamers, the engineers are ordered to blow out every two hours. But it is more usual to do so only once a day.

This method, however, of blowing out furnishes but a partial remedy for the evils we have alluded to : a loose deposite will perhaps be removed by such means, but an incrustation, more or less according to the circumstances and quality of the water, will be formed; besides which, the temptation to work the vessel with efficiency for the moment influences the engine-men to neglect blowing out; and it is found that this class of persons can rarely be relied upon to resort to this remedy with that constancy and regularity which are essential for the due preservation of the boilers. The class of steam vessels which, at present, are exposed to the greatest injury from these causes are the sea-going steamers employed by the Admiralty; and we find, by a report made by Messrs. Lloyd and Kingston to the Admiralty, in August, 1834, that it is admitted that the method of blowing out is, even when daily attended to, ineffectual. "The water in the boiler," these gentlemen observe, "is kept from exceeding a certain degree of saltness, by periodically blowing a portion of it into the sea; but whatever care is taken, in long voyages especially, salt will accumulate, and sometimes in great quantities and of great hardness, so that it is with difficulty it can be removed. Boilers are thus often injured as much in a few months as they would otherwise be in as many years. The other evil necessarily resulting from this state of things is, besides the rapid destruction of the boilers, a great waste of fuel, occasioned by the difficulty with which the heat passes through the incrustation on the inside, by the leaks which are thereby caused, and by the practice of blowing out periodically, as before mentioned, a considerable portion of the boiling water."

It would be impracticable to carry on board the vessel a sufficient quantity of pure fresh water to work the engine exclusively by its means. To accomplish this, it would be necessary to have a sufficient supply of cold water to keep the condensing cistern cold, to supply the jet in the condenser, and to have a reservoir in which the warm water coming from the waste pipe of the cold cistern might be allowed to cool. Engineers have therefore directed their attention to some method by which the steam may be condensed without a jet, and after condensation be preserved for the purpose of feeding the boiler. If this could be accomplished, it would not be necessary to provide a greater quantity of pure water than would be sufficient to make up the small portion of waste which might proceed from leakage and from other causes; and it is evident that this portion might always be readily obtained by the distillation of sea water, which might be effected by a small vessel exposed to the same fire which acts upon the boiler.

(115.) Mr. *Samuel Hall*, of *Basford*, near Nottingham, has taken out patents for a new form of condenser, contrived for the attainment of these ends, besides some other improvements in the engine.

The condenser of Mr. *Hall* consists of a great number of narrow tubes immersed in a cistern of cold water: the steam as it passes from the cylinder, after having worked the piston, enters these tubes, and is immediately condensed by their cold surfaces. It flows in the form of water from their remote extremities, and is drawn off by the air-pump, and conducted in the usual way to a cistern from which the boiler is fed. In the marine engines constructed under Mr. *Hall's* patents, the tubes of the condenser being in an upright or vertical position, the steam flows from the cylinder into the upper part of the condenser, which is a low flat chamber, in the bottom of which is inserted the upper extremities of the tubes, through which the steam passes downward, and as it passes is condensed. It flows thence into a similar chamber below, from whence it is drawn off by the air-pump.

It is evident that at sea an unlimited supply of cold water

may be obtained to keep the condensing cistern cold, so that a perfect condensation may always be effected by these tubes, if they be made sufficiently small. The water formed by the condensed steam will be pure distilled water; and if the boiler be originally filled with water which does not hold in solution any earthy or other matter which might be deposited or incrusted, it may be worked for any length of time without injury. The small quantity of waste from leakage is supplied in Mr. Hall's engine by a simple apparatus in which a sufficient quantity of sea-water may be distilled.

The following are the advantages, as stated by Mr. Hall, to be gained by his condenser:—

1. A saving of fuel, amounting in some cases to so much as a third of the ordinary consumption.

2. The preservation of the boilers from the destruction produced in common engines by the corrosive action of sea or other impure water, and by incrustations of earthy matter.

3. The saving of the time spent in cleaning the boilers.

4. A considerable increase of power, owing to the cleanness of the boilers; the absence of injected water to be pumped out of a vacuum; the greater perfection of the vacuum; the better preservation of the piston and valves of the air-pump; and (by another contrivance of his) the more perfect lubrication of the parts of the engine.

5. The water in the boiler being constantly maintained at the same height by self-acting arrangement.

6. The size of a boiler exerting a given power, being much smaller than the common kind, owing to its more perfect action.

Messrs. *Lloyd* and *Kingston* were employed by government to examine and report the effects of Mr. Hall's boilers, and they stated in their report, already referred to, that the result is so succcessful as to leave nothing to be wished for. Among the advantages which they enumerate are the increased durability of the engines; the prevention of accidents through carelessness, or otherwise, arising from the condenser and air-pump becoming choked with injection water; and the additional security against the boilers being burnt in consequence of the water being suffered to get too low. But the greatest advantages, compared with which they consider all others to be of secondary importance, are the increased durability of the boilers and the saving of fuel.

About sixteen engines, built either wholly upon Mr. Hall's principle, or having his condenser attached to them, have now (October, 1835) been working in different parts of England, and on board different vessels for various periods, from three years to three months; and it appears from the concurrent testimony of the proprietors and managers of them, that they are attended with all the advantages which the patentee engaged for. The part of the contrivance the performance of which would have appeared most doubtful would have been the maintenance of a sufficiently good vacuum in the condenser, in the absence of the usual method of condensation by the injection of cold water; nevertheless it appears that a better vacuum is sustained in these engines than in the ordinary engines which condense by jet. The barometer gauge varies from 29 to 29½ inches, and in some cases comes up to 30 inches, according to the state of the barometer: this is a vacuum very nearly perfect, and indeed may be said to be so for all practical purposes. The *Prince Llewellyn* and the *Air* steam packets, belonging to the St. George Steam Packet Company, have worked such a pair of these engines for about a year. The *City of London* steam packet, the property of the General Steam Navigation Company, has been furnished with two fifty-horse engines, and has worked them during the same period. In all cases the boilers have been found perfectly free from scale or incrustation; and the deposite is either absolutely nothing or very trifling, requiring the boiler to be swept about once in half a year, and sometimes not so often. The trial which has been made of these engines in the navy has proved satisfactory, so far as it has been carried. The Lords of the Admiralty have lately ordered a pair of seventy-horse engines to be constructed on this principle for a vessel now (October, 1835) in process of construction;* and another vessel in all respects similar, except having copper boilers, is likewise ordered; so that a just comparison may be made. It would, however, have been more fair if both vessels had been provided with iron boilers, since copper does not receive incrustation as readily as iron.

It would seem that the advantages of these boilers in the vessels of the St. George Steam Packet Company were regarded by the directors as sufficiently evident, since, after more than a year's experience, they are about to place a pair of ninety-horse engines of this kind in a new and powerful steamer called the *Hercules*.

Engines furnished with Mr. Hall's apparatus have not yet, so far as I am informed, been tried with reference to the power exerted by the consumption of a given quantity of fuel. The mere fact of a good vacuum being sustained in the condenser cannot by itself be regarded as a conclusive proof of the efficiency of the engine, without the water or air introduced by a condensing jet. Mr. Hall, nevertheless, uses as large an air-pump as that of an ordinary condensing engine, and recommends even a larger one. For what purpose, it may be asked, is such an appendage introduced? If there be nothing to be removed but the condensed steam, a very small pump ought to be sufficient. It is not wonderful that a good vacuum is sustained in the condenser, if the power expended on the air-pump is employed in *pumping away uncondensed steam.* Such a contrivance would be merely a deception, giving an apparent but no real advantage to the engine.

* This order has, as Mr. Hall informs me, been given without requiring any guarantee as to the performance of the engines.

Having mentioned these advantages, which are said to arise from Mr. Hall's condenser, it is right to state that it is in fact a reproduction of an early invention of Mr. Watt. There is in possession of James Watt, Esquire, a drawing of a condenser laid before parliament in 1776, in which the same method of condensing without a jet is proposed. Mr. Watt, however, finding that he could not procure by that means so sudden or so perfect a vacuum as by injection, abandoned it. I believe he also found that the tubes of the condenser became furred with a deposite which impeded the process of condensation. It would seem, however, that Mr. Hall has found means to obviate these effects. It is right to add, that Mr. Hall, in his specification, distinctly disclaims all claim to the method of condensing by tubes without jet.

There is another part of Mr. Hall's contrivance which merits notice. In all engines, a considerable quantity of steam is allowed to escape from the safety valve. Whenever the vessel stops, the steam, which would otherwise be taken from the boiler by the cylinders, passes out through this valve into the atmosphere. Also, whenever the cylinders work at under-power, and do not consume the steam as fast as it is produced by the boiler, the surplus steam escapes through the valve. Now, according to the principle of Mr. Hall's method, it is necessary to save the water which thus escapes in vapour, since otherwise the pure water of the boiler would be more rapidly wasted. Mr. Hall accordingly places a safety valve of peculiar construction in communication with a tube which leads to the condenser, so that whenever, either by stopping the engine or diminishing its working power, steam accumulates in the boiler, its increased pressure opens the safety valve, and it passes through this pipe to the condenser, where it is reconverted into water, and pumped off by the air-pump into the cistern from which the boiler is fed.

The attainment of an object so advantageous as to extend

the powers of steam navigation, and to render the performance of voyages of any length practicable, so far as the efficiency of the machinery is concerned, has naturally stimulated the inventive genius of the country. The preservation of the boiler by the prevention of deposite and incrustation is an object of paramount importance; and its attainment necessarily involves, to a certain degree, another condition on which the extension of steam voyages must depend, viz. the economy of fuel. In proportion as the economy of fuel is increased, in the same proportion will the limit to which steam navigation may be carried be extended.

(116.) A patent has been obtained by Mr. Thomas Howard of London for a form of engine possessing much novelty and ingenuity, and having pretensions to the attainment of a very extraordinary economy of fuel, in addition to those advantages which have been already explained as attending Mr. Hall's engines. In these engines, as in Mr. Hall's, the steam is constantly reproduced from the same water, so that pure or distilled water may be used ; but Mr. Howard dispenses with the use of a boiler altogether. The steam also with which he works is in a state essentially different from the steam used in ordinary engines. In these, the vapour is raised directly from the water in a boiling state, and it contains as much water as it is capable of holding at its temperature. Thus, at the temperature of 212°, a cubic foot of steam used in common engines will contain about a cubic inch of water; but in the contrivance of Mr. Howard, a considerable quantity of heat is imparted to the steam before it passes into the cylinder in addition to what is necessary to maintain it in the vaporous form.

A quantity of mercury is placed in a shallow wrought-iron vessel over a coke fire, by which it is maintained at the temperature of from 400° to 500°. The surface exposed to the fire is three-fourths of a square foot for each horse power. The upper surface of the mercury is covered by a very thin plate of iron, which rests in contact with it, and which is so contrived as to present about four times as much surface as that exposed beneath to the fire. Adjacent to this a vessel of water is placed, kept heated nearly to the boiling point, which communicates by a nozle and valve with the chamber or vessel immediately above the mercury. At intervals corresponding to the motion of the piston, a small quantity of water is injected from this vessel, and thrown upon the plate of iron which rests upon the hot mercury: from this it receives the heat necessary not only to convert it into steam, but to expand that steam, and raise it to a temperature above the temperature it would receive if raised in immediate contact with water. In fact, the steam thus produced will have a temperature not corresponding to its pressure, but considerably above that point, and it will therefore be in circumstances under which it will part with more or less of its heat, and allow its temperature to be lowered without being even partially condensed, whereas steam used in the ordinary steam engines must be more or less condensed by the slightest diminution of its temperature. The quantity of liquid injected into the steam chamber must be regulated by the power at which the engine is intended to work. The fire is supplied with air by a blowing machine, which is subject to exact regulation. The steam, produced in the manner already explained, passes into a chamber which surrounds the working cylinder; and this chamber itself is enclosed by another space, through which the air from the furnace must pass before it reaches the flue. In this way it imparts its redundant heat to the steam which is about to work the cylinder, and raises it to a temperature of about 400°; the pressure, however, not exceeding 25 lbs. per square inch. The arrangement of valves for the admission of the steam to the cylinder is such as to cause the steam to act expansively.

The vacuum on the opposite side of the piston is maintained by condensation in the following manner:—The condenser is a copper vessel placed in a cistern constantly supplied with cold water, and the steam flows to it from the cylinder by an eduction pipe in the usual way: a jet is admitted to it from an adjacent vessel, which, before the engine commences work, is filled with distilled water; the condensing water and condensed steam are pumped from the condenser by air-pumps of the usual construction, but smaller, inasmuch as there is no air to be withdrawn, as in common engines. The warm water thus pumped out of the condenser is driven into a copper pipe or worm, which is carried with many coils through a cistern of cold water, so that when it arrives at the end of this pipe it is reduced to the common temperature of the atmosphere. The pipe is then conducted into the vessel of distilled water already mentioned, and the water flowing from it continually replaces the water which flows into the condenser through the condensing jet. The condensing water being purged of air, a very small air-pump is sufficient; since it has only to exhaust the condenser and tubes at starting, and to remove whatever air may enter by casual leakage. The patentee states that the condensation takes place as rapidly and as perfectly as in the best steam engine, and it is evident that this method of condensation is applicable even where the mercurial generator already described may not be employed. The vessel from which the water is injected into the mercurial generator is likewise fed by the air-pump connected with the condenser. There is another pipe besides the copper worm already described, which is carried from the hot well to this vessel, and the water is of course returned through it without being cooled. This vessel is likewise sufficiently exposed to the action of the fire to maintain it at a temperature somewhat below the boiling point.

An apparatus of this construction was in the spring of the present year (1835) placed in the Admiralty steamer called the Comet, in connexion with a pair of forty-horse engines The patentee states that these engines were ill adapted to the contrivance; nevertheless, the vessel was successfully worked in the Thames for 800 miles: she also performed a voyage from Falmouth to Lisbon, but was prevented from returning by an accident which occurred to the machinery near the latter port. In this experimental voyage, the consumption of fuel is stated never to have exceeded a third of her former consumption, when worked by Bolton and Watt's engines; the former consumption of coals being about 800 lbs. per hour, and the consumption with Mr. Howard's engine being under 250 lbs. of coke per hour.

After this failure (which, however, was admitted to be one of accident and not of principle) the government did not consider itself justified in bestowing further time or incurring greater expense in trying this engine. Mr. Howard, however, has himself built a new vessel, in which he is about to place a pair of forty-horse engines. This vessel is now (December, 1835) nearly ready, and will bring the question to issue by a fair experiment. The advantages of the contrivance as enumerated by the patentee are:—

*First*, The small space and weight occupied by the machinery, arising from the absence of a boiler.

*Secondly*, The diminished consumption of fuel.

*Thirdly*, The reduced size of the flues.

*Fourthly*, The removal of the injurious effects arising from deposite and incrustation.

*Fifthly*, The absence of smoke.

Some of these improvements, if realized, will be attended with important advantages in steam navigation. Steamers of a given tonnage and power will have more disposable space for lading and fuel, and in short voyages may carry greater freight, or an increased number of passengers; or by taking a larger quantity of fuel,* may make greater runs

* The fuel used in this form of engine is coke, and not coal. A ton of coke occupies the same space as two tons of coal; the saving of tonnage, therefore,

than are now attainable ; or, finally, with the same tonnage and the same lading, they may be supplied with more powerful machinery.

(117.) To obtain from the moving power its full amount of mechanical effect in propelling the vessel, it would be necessary that its force should propel, by constantly acting against the water in a horizontal direction, and with a motion contrary to the course of the vessel. No system of mechanical propellers has, however, yet been contrived capable of perfectly attaining this end. Patents have been granted for many ingenious mechanical combinations to impart to the propelling surfaces such angles as appeared to the respective contrivers most advantageous. In most of these, however, the mechanical complexity has formed a fatal objection. No part of the machinery of a steam vessel is so liable to become deranged at sea as the paddle wheels ; and, therefore, such simplicity of construction as is compatible with those repairs which are possible on such emergencies is quite essential for safe practical use.

The ordinary paddle wheel, as I have already stated, is a wheel revolving upon a shaft driven by the engine, and carrying upon its circumference a number of flat boards, called paddle boards, which are secured by nuts or braces in a fixed position ; and that position is such that the planes of the paddle boards diverge nearly from the centre of the shaft on which the wheel turns. The consequence of this arrangement is that each paddle board can only act in that direction which is most advantageous for the propulsion of the vessel when it arrives near the lowest point of the wheel. In

by the increased economy of fuel will not be in so great a proportion as the saving of fuel. A quantity of fuel of equivalent power will occupy about half the present space, but the displacement or immersion which it produces will be only one-fourth of its present effect.

*Fig.* 67.

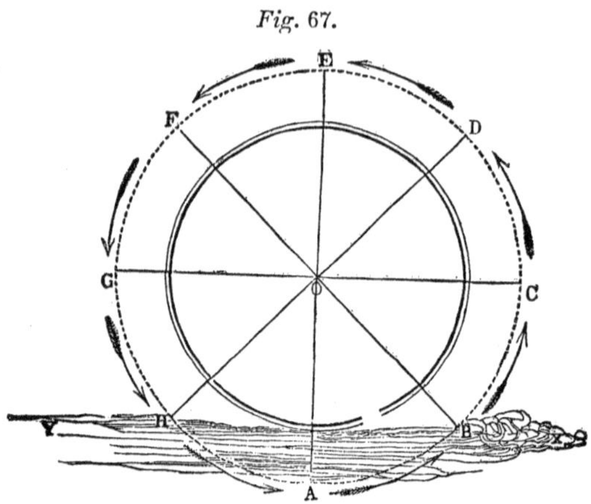


figure 67, let O be the shaft on which the common paddle wheel revolves; the position of the paddle boards are represented at A, B, C, &c.; X, Y represents the water line, the course of the vessel being supposed to be from X to Y; the arrows represent the direction in which the paddle wheel revolves. The wheel is immersed to the depth of the lowest paddle board, since a less degree of immersion would render a portion of the surface of each paddle board mechanically useless. In the position A the whole force of the paddle board is efficient for propelling the vessel; but, as the paddle enters the water in the position H, its action upon the water, not being horizontal, is only partially effective for propulsion: a part of the force which drives the paddle is expended in depressing the water, and the remainder in driving it contrary to the course of the vessel, and, therefore, by its reaction producing a certain propelling effect. The tendency, however, of the paddle entering the water at H, is to form a hollow or trough, which the water, by its ordinary property, has a continual tendency to fill up. After passing the lowest point A, as the paddle approaches the position B, where it emerges from the water, its action again becomes oblique, a part only having a propelling effect, and the remainder having a tendency to raise the water, and throw up a wave and spray behind the paddle wheel. It is evident that the more deeply the paddle wheel becomes immersed the greater will be the proportion of the propelling power thus wasted in elevating and depressing the water; and, if the wheel were immersed to its axis, the whole force of the paddle boards, on entering and leaving the water, would be lost, no part of it having a tendency to propel. If a still deeper immersion takes place, the paddle boards above the axis would have a tendency to retard the course of the vessel. When the vessel is, therefore, in proper trim, the immersion should not exceed nor fall short of the depth of the lowest paddle ; but for various reasons it is impossible in practice to maintain this fixed immersion : the agitation of the surface of the sea, causing the vessel to roll, will necessarily produce a great variation in the immersion of the paddle wheels, one becoming frequently immersed to its axle, while the other is raised altogether out of the water. Also the draught of water of the vessel is liable to change, by the variation in her cargo : this will necessarily happen in steamers which take long voyages. At starting they are heavily laden with fuel, which as they proceed is gradually consumed, whereby the vessel is lightened; and it does not appear that it is practicable to use sea water as ballast to restore the proper degree of immersion.

(118.) Among the contrivances which have been proposed for remedying these defects of the common paddle wheel by introducing paddle boards capable of shifting their position as they revolve with the circumference of the wheel, the only one which has been adopted to any considerable extent in practice, is that which is commonly known as Morgan's Paddle Wheel. The original patent for this contrivance was granted to Elijah Galloway, and sold by him to Mr. William Morgan. Subsequently to the purchase some improvements in its structure and arrangements were introduced, and it is now extensively adopted by government in the Admiralty steamers. It was first tried on board His Majesty's steamer the CONFIANCE; and after several successful experiments was ordered by the Lords of the Admiralty to be introduced on board the FLAMER, the FIREBRAND, the COLUMBIA, the SPITFIRE, the LIGHTNING, a large war steamer called the MEDEA,* the TARTARUS, the BLAZER, &c. It has been tried by government in several well-conducted experiments, where two vessels of precisely the same model, supplied with similar engines of equal power, and propelled, one by Morgan's paddle wheels, and the other by the common paddle wheels; when it was found that the advantage of the former, whether in smooth or in rough water, was quite apparent. One of the commanders in these experiments (Lieutenant Belson) states that the improvement in the speed of the CONFIANCE, after being supplied with these wheels, was proportionately greater in a sea way than in smooth water; that their action was not impeded by the waves, since the variation of the velocity of the engine did not exceed one or two revolutions per minute: the vessel's way was never stopped, and there was no sensible increase of vibration on the paddle boxes during the gale. Another commander reported that on a comparison of the CONFIANCE and a similar and equally powerful vessel, the CARRON, the CONFIANCE performed in fifty-four hours the voyage which occupied the CARRON eighty-four hours in running. Independently of the great saving of fuel effected, (namely, ten bushels per hour,†) or the time saved in running the same distance, other advantages have been secured

* This splendid ship is 860 tons burden, with engines of 220 horse-power.

† See Report quoted in Mechanic's Magazine, vol. xxii. p. 275. This saving cannot amount to less than 40 per cent. upon the whole consumption of fuel; it certainly is considerably beyond what I should have conceived to be possible; but I have no reason to doubt the accuracy of the Report. I estimate the former consumption at 10 pounds per horse-power per hour, which on 220 horse-power would be 2200 pounds, of which 840 pounds = 10 bushels, would be about 4-10ths.

y the modification in question. On a comparison of the respective logs of the two vessels, it appeared that the CONFIANCE had gained by the alteration in her wheels an increase of speed amounting to 2 knots on 7 in smooth water, and 2½ knots on 4 to 4½ knots in rough weather; that the action of the paddles did not bring up the engine or retard their velocity in a head sea; that in rolling their action assisted in righting the vessel; and that the wear and strain, as well on the vessel as on the engines, were materially reduced. With respect to the durability of these wheels, the commander of the FLAMER reported in January, 1834, that in six weeks of the most tempestuous weather they found them to act remarkably well, without even a single float being shifted.*

This paddle wheel is represented in fig. 68. The contrivance may be shortly stated to consist in causing the wheel which bears the paddles to revolve on one centre, and the radial arms which move the paddles to revolve on another centre. Let A B C D E F G H I be the polygonal circumference of the paddle wheel, formed of straight bars, securely connected together at the extremities of the spokes or radii of the wheel which turns on the shaft which is worked by the engine; the centre of this wheel being at O. So far this wheel is similar to the common paddle wheel; but the paddle boards are not, as in the common wheel, fixed at A B C, &c., so as to be always directed to the centre O, but are so placed that they are capable of turning on axles which are always horizontal, so that they can take any angle with respect to the water which may be given to them. From the centres, or the line joining the pivots on which these paddle boards turn, there proceed short arms K, firmly fixed to the

* See a more detailed account of these reports in the Mechanic's Magazine, vol. xxii. page 274, from which I have taken the drawing of this paddle wheel; and also see Report of the Committee of the House of Commons on Steam Navigation to India, Evidence of William Morgan age 95.

paddle boards at an angle of about 120°. On a motion given to this arm K, it will therefore give a corresponding angular motion to the paddle board, so as to make it turn on its pivots. At the extremities of the several arms marked K is a pin or pivot, to which the extremities of the radial arms L are severally attached, so that the angle between each radial arm L and the short paddle arm K is capable of being changed by any motion imparted to L; the radial arms L are connected at the other end with a centre P, round which they are capable of revolving. Now since the points A B C, &c., which are the pivots on which the paddle boards turn, are moved in the circumference of a circle, of which the centre is O, they are always at the same distance from that point; consequently they will continually vary their distance from the other centre P. Thus, when a paddle board arrives at that point of its revolution at which the centre P lies precisely between it and the centre O, its distance from P is less than in any other position. As it departs from that point, its distance from the centre P gradually increases until it arrives at the opposite point of its revolution, where the centre O is exactly between it and the centre P; then the distance of the paddle board from the centre P is greatest. This constant change of distance between each paddle board and the centre P is accommodated by the variation of the angle between the radial arm L and the short paddle board arm K; as the paddle board approaches the centre P this gradually diminishes; and as the distance of the paddle board from P increases, the angle is likewise augmented. This change in the magnitude of the angle, which thus accommodates the varying position of the paddle board with respect to the centre P, will be observed in the figure. The paddle board D is nearest to P; and it will be observed that the angle contained between L and K is there very acute; at E the angle between L and K increases, but is still acute; at F it increases to a right angle; at G it becomes obtuse; and at K, where it is most distant from the centre P, it becomes most obtuse. It again diminishes at I, and becomes a right angle between A and B. Now this continual shifting of the direction of the short arm K is necessarily accompanied by an equivalent change of position in the paddle board to which it is attached; and the position of the second centre P is, or may be, so adjusted that this paddle board, as it enters the water and emerges from it, shall be such as shall be most advantageous for propelling the vessel, and therefore attended with less of that vibration which arises chiefly from the alternate depression and elevation of the water, owing to the oblique action of the paddle boards.*

(*i*) The relative value of the two wheels, namely, the common paddle wheel and that of Morgan, has been investigated by Professor Barlow of the Military School at Woolwich, and the results published in a paper of much ability in the Philosophical Transactions for 1834. By this paper it appears, that, when the paddles are not wholly immersed, the wheel of Morgan has no important advantage over the other, and only acquires one when the wheel wallows. But the most important of his inferences is, that the common paddle is least efficient when in a vertical position, contrary to the usual opinion. From this we have a right to infer that the search for a form of wheel which shall always keep the paddle vertical is one whose success need not be attended with any important consequence. The superior qualities of Morgan's wheel when the paddles are deeply immersed is ascribed by Barlow to the lessening of the shock sustained by the common paddle wheel when it strikes the water. This being the case, the triple wheel of Stevens is probably superior to that of Morgan in its efficiency, while it has the advantage of being far simpler and less liable to be put out of order.—A. E.

* A paddle wheel resembling this has lately been constructed by Messrs. Seawards. It has been charged by Mr. Morgan as being a colourable invasion of his patent, and the dispute has been brought into the courts of law.

(119.) To form an approximate estimate of the limit of the present powers of steam navigation, it will be necessary to consider the mutual relation of the capacity or tonnage of the vessel; the magnitude, weight, and power of the machinery; the available stowage for fuel; and the average speed attainable in all weathers, as well as the general purposes to which the vessel is to be appropriated, whether for the transport of goods and merchandise, or merely of despatches and passengers. That portion of the capacity of the vessel which is appropriated to the moving power consists of the space occupied by the machinery and the space occupied by the fuel; the magnitude of the latter will necessarily depend upon the length of the voyage which the vessel must make without receiving a fresh supply of coals. If the voyage be short, this space may be proportionally limited, and a greater portion of room will be left for the machinery. If, on the contrary, the voyage be longer, a greater stock of coals will be necessary, and a less space will remain for the machinery. More powerful vessels, therefore, in proportion to their tonnage, may be used for short than for long voyages.

Taking an average of fifty-one voyages made by the Admiralty steamers, from Falmouth to Corfu and back during four years ending June, 1834, it was found that the average rate of steaming, exclusive of stoppages, was $7\frac{1}{4}$ miles per hour, taken in a direct line between the places, and without allowing for the necessary deviations in the course of the vessel. The vessels which performed this voyage varied from 350 to 700 tons burthen by measurement, and were provided with engines varying from 100 horse to 200 horse-power, with stowage for coals varying from 80 to 240 tons. The proportion of the power to the tonnage varied from 1 horse to 3 tons to 1 horse to 4 tons; thus, the MESSENGER had a power of 200 horses, and measured 730 tons; the FLAMER had a power of 120 horses.

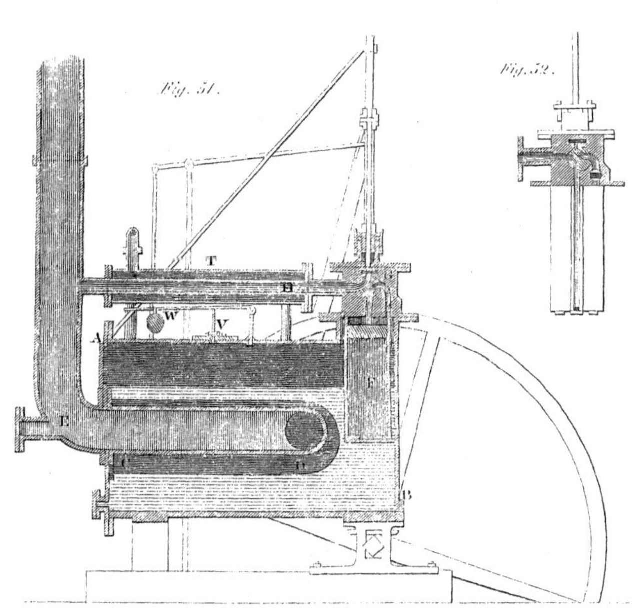


Fig. 51.

Fig. 52.

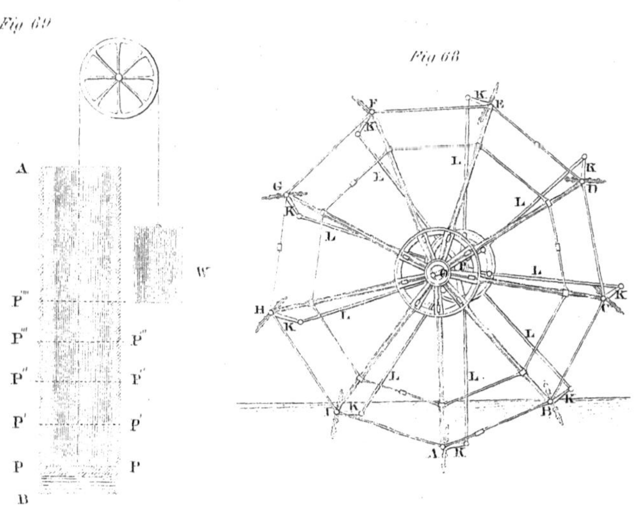


Fig 69

Fig 68

and measured 500 tons; the COLUMBIA had 120 horses, and measured 360 tons.

In general, it may be assumed that for the shortest class of trips, such as those of the Margate steamers, and the packets between Liverpool or Holyhead and Dublin, the proportion of the power to the tonnage should be that of 1 horse-power to every 2 tons by measure; while for the longest voyages the proportion would be reduced to 1 horse to 4 tons, voyages of intermediate lengths having every variety of intermediate proportion.

Steamers thus proportioned in their power and tonnage may then, on an average of weathers, be expected to make $7\frac{1}{4}$ miles an hour while steaming, which is equivalent to 174 miles per day of twenty-four hours. But, in very long voyages, it rarely happens that a steamer can work constantly without interruption. Besides stress of weather, in which she must sometimes lie-to, she is liable to occasional derangements of her machinery, and more especially of her paddles. In almost every long voyage hitherto attempted, some time has been lost in occasional repairs of this nature while at sea. We shall perhaps, therefore, for long voyages, arrive at a more correct estimate of the daily run of a steamer by taking it at 160 miles.*

By a series of carefully conducted experiments on the consumption of coals, under marine boilers and common land boilers, which have been lately made at the works of Mr. Watt, near Birmingham, it has been proved that the

* The American reader will hardly be able to refrain from a smile at this estimate of Dr. Lardner of the speed of steamboats, founded upon the most improved practice of Europe at the close of the year 1835. The boats on the Hudson river have for years past averaged a speed of 15 miles per hour, and the Lexington, which was constructed for a navigation part of which is performed in the open ocean, could probably keep up a speed of the same amount, except in severe storms. With boats, constructed on the principles of those which navigate Long Island Sound and the Chesapeake, we should not fear to assume 12 miles per hour, or upward of 280 miles per day, as their average rate of crossing the ocean.—A. E.

consumption of fuel under marine boilers is less than under land boilers, in the proportion of 2 to 3 very nearly. On the other hand, I have ascertained from general observation throughout the manufacturing districts in the North of England, that the average consumption of coals under land boilers of all powers above the very smallest class is at the rate of 15 pounds of coals per horse-power per hour. From this result, the accuracy of which may be fully relied upon, combined with the result of the experiments just mentioned at Soho, we may conclude that the average consumption of marine boilers will be at the rate of 10 lbs. of coal per horse-power per hour. Mr. Field, of the firm of Maudslay and Field, in his evidence before a Select Committee of the House of Commons on Steam Navigation to India, has stated from his observation, and from experiments made at different periods, that the consumption is only 8 lbs. per horse-power per hour. In the evidence of Mr. William Morgan, however, before the same committee, the actual consumption of fuel on board the Mediterranean packets is estimated at 16 cwt. per hour for engines of 200 horse-power, and 8¼ cwt. for engines of 100 horse-power. From my own observation, which has been rather extensive both with respect to land and marine boilers, I feel assured that 10 lbs. per hour more nearly represents the practical consumption than the lower estimate of Mr. Field. We may then assume the daily consumption of coal by marine boilers, allowing them to work upon an average for 22 hours, the remainder of the time being left for casual stoppages, at 220 lbs. of coal per horse-power, or very nearly 1 ton for every ten horses' power. In short voyages, where there will be no stoppage, the daily consumption will a little exceed this ; but the distance traversed will be proportionally greater.

When the proportion of the power to the tonnage remains unaltered, the speed of the vessel does not materially change. We may therefore assume that ten pounds of coal per horse power will carry a sea-going steamer adapted for long voyages $7\frac{1}{4}$ miles direct distance ; and therefore to carry her 100 miles will require 138 pounds, or the $\frac{1}{16}$th part of a ton nearly. Now the Mediterranean steamers are capable of taking a quantity of fuel at the rate of $1\frac{1}{4}$ tons per horse-power ; but the proportion of their power to their tonnage is greater than that which would probably be adapted for longer runs. We shall, therefore, perhaps be warranted in assuming that it is practicable to construct a steamer capable of taking $1\frac{1}{2}$ tons of fuel per horse power. At the rate of consumption just mentioned, this would be sufficient to carry her 2400 miles in average weather; but as an allowance of fuel must always be made for emergencies, we cannot suppose it possible for her to encounter this extreme run. Allowing, then, spare fuel to the extent of a quarter of a ton per horse-power, we should have as an extreme limit of a steamer's practicable voyage, without receiving a relay of coals, a run of about 2000 miles.

(120.) This computation is founded upon results obtained from the use chiefly of the North of England coal. It has, however, been stated in evidence before the select committee above mentioned, that the *Llangennech* coal of Wales is considerably more powerful. Captain Wilson, who commanded the HUGH LINDSAY steamer in India, has stated that this coal is more powerful than Newcastle, in the proportion of 9 to $6\frac{1}{2}$.* Some of the commanders of the Mediterranean packets have likewise stated that the strength of this coal is greater than that of Newcastle in the proportion of 16 to 11.† So far then as relates to this coal, the above estimate must be modified, by reducing the consumption nearly in the proportion of 3 to 2.

The class of vessels best fitted for undertaking long voyages, without relays of coal, would be one from about 800 to

* Vide Report of Select Committee on Steam Navigation, p. 152.

† Ibid. p. 7.

1000 tons measurement furnished with engines from 200 to 250 horse-power.* Such vessels could take a supply of from 300 to 400 tons of coals, which, being consumed at the rate of from 20 to 25 tons per day, would last about fifteen days.

Applying these results, however, to particular cases, it will be necessary to remember that they are average calculations, and must be subject to such modifications as the circumstances may suggest in the particular instances : thus, if a voyage is contemplated under circumstances in which an adverse wind generally prevails, less than the average speed must be allowed, or, what is the same, a greater consumption of fuel for a given distance. Against a strong head wind, in which a sailing vessel would double-reef her top-sails, even a powerful steamer cannot make more than from 2 to 3 miles an hour, especially if she has a head sea to encounter.

(121.) In considering the general economy of fuel, it may be right to state, that the results of experience obtained in the steam navigation of our channels, and particularly in the case of the Post-office packets on the Liverpool station, have clearly established the fact, that by increasing the ratio of the power to the tonnage, an actual saving of fuel in a given distance is effected, while at the same time the speed of the vessel is increased. In the case of the Post-office steamers called the DOLPHIN and the THETIS, (Liverpool station,) the power has been successively increased, and the speed proportionably augmented ; but the consumption of fuel per voyage between Liverpool and Dublin has been diminished. This, at first view, appears inconsistent with the known theory of the resistance of solids moving through fluids; since this resistance increases in the same proportion as the square of the speed. But this physical principle is founded

* Engines in steam vessels generally work considerably above their nominal power. The power, however, to which we uniformly refer is the nominal power, or that power at which they would work with steam of the ordinary pressure.

on the supposition that the immersed part of the floating body remains the same. Now I have myself proved by experiments on canals, that when the speed of the boat is increased beyond a certain limit, its draught of water is rapidly diminished; and in the case of a large steam raft constructed upon the river Hudson, it was found that when the speed was raised to 20 miles an hour, the draught of water was diminished by 7 inches. I have therefore no doubt that the increased speed of steamers is attended with a like effect; that, in fact, they rise out of the water; so that, although the resistance is increased by reason of their increased speed, it is diminished in a still greater proportion by reason of their diminished immersion.

Meanwhile, whatever be the cause, it is quite certain that the resistance in moving through the water must be diminished, because the moving power is always in proportion to the quantity of coals consumed, and at the same time in the proportion to the resistance overcome. Since, then, the quantity of coals consumed in a given distance is diminished while the speed is increased, the resistance encountered throughout the same distance must be proportionally diminished.

(122.) Increased facility in the extension and application of steam navigation is expected to arise from the substitution of iron for wood, in the construction of vessels. Hitherto iron steamers have been chiefly confined to river navigation; but there appears no sufficient reason why their use should be thus limited. For sea voyages they offer many advantages; they are not half the weight of vessels of equal tonnage constructed of wood; and, consequently, with the same tonnage they will have less draught of water, and therefore less resistance to the propelling power; or, with the same draught of water and the same resistance, they will carry a proportionally heavier cargo. The nature of their material renders them more stiff and unyielding than timber; and they do not suffer that effect which is called *hog ging*, which arises from a slight alteration which takes place in the figure of a timber vessel in rolling, accompanied by an alternate opening and closing of the seams. Iron vessels have the further advantage of being more proof against fracture upon rocks. If a timber vessel strike, a plank is broken, and a chasm opened in her many times greater than the point of rock which produces the concussion. If an iron vessel strike, she will either merely receive a dinge, or be pierced by a hole equal in size to the point of rock which she encounters. Some examples of the strength of iron vessels was given by Mr. Macgregor Laird, in his evidence before the Committee of the Commons on Steam Navigation, among which the following may be mentioned :—An iron vessel, called the ALBURKAH, in one of their experimental trials got aground, and lay upon her anchor : in a wooden vessel the anchor would probably have pierced her bottom ; in this case, however, the bottom was only dinged. An iron vessel, built for the Irish Inland Navigation Company, was being towed across Lough Derg in a gale of wind, when the towing rope broke, and she was driven upon rocks, on which she bumped for a considerable time without any injury. A wooden vessel would in this case have gone to pieces. A further advantage of iron vessels (which in warm climates is deserving of consideration) is their greater coolness and perfect freedom from vermin.

(123.) The greatest speed which has yet been attained upon water by the application of steam has been accomplished in the case of a river steamer of peculiar form, which has been constructed upon the river Hudson. This boat, or rather raft, consisted of two hollow vessels formed of thin sheet iron, somewhat in the shape of spindles or cigars, (from whence it was called the *cigar boat.*) In the thickest part these floats were eight feet in diameter, tapering toward the ends, and about 300 feet long : these floats or buoys, being placed parallel to each other, having a distance of more than 16 feet between them, supported a deck or raft 300 feet long, and 32 feet wide. A paddle wheel 30 feet in diameter and 16 feet broad revolved between the spindles, impelled by a steam engine placed upon the deck. This vessel drew about 30 inches of water, and attained a speed of from 20 to 25 miles an hour: she ran upon a bank in the river Hudson, and was lost. The projector is now employed in constructing another vessel of still larger dimensions. It is evident that such a structure is altogether unfitted for sea navigation. In the case of a wide navigable river, however, such as the Hudson, it will no doubt be attended with the advantage of greater expedition.

(124.) Several projects for the extension of steam navigation to voyages of considerable length have lately been entertained both by the public and by the legislature, and have imparted to every attempt to improve steam navigation increased interest. A committee of the House of Commons collected evidence and made a report in the last session in favour of an experiment to establish a line of steam communication between Great Britain and India. Two routes have been suggested by the committee, each being a continuation of the line of Admiralty steam packets already established to Malta and the Ionian Isles. One of the routes proposed is through Egypt, the Red Sea, and across the Indian Ocean to Bombay, or some of the other Presidencies; the other across the north part of Syria to the banks of the Euphrates, by that river to the Persian Gulf, and from thence to Bombay. Each of these routes will be attended with peculiar difficulties, and in both a long sea voyage will be encountered.

In the route by the Red Sea, it is proposed to establish steamers between Malta and Alexandria, (860 miles.) A steamer of 400 tons burthen and 100 horse-power would perform this voyage, upon an average of all weathers incident to the situation, in from 5 to 6 days, consuming 10 tons of coal per day. But it is probable that it might be found more advantageous to establish a higher ratio between the power and the tonnage. From Alexandria, the transit might be effected by land across the Isthmus to Suez—a journey of from 4 to 5 days—by caravan and camels; or the transit might be made either by land or water from Alexandria to Cairo, a distance of 173 miles; and from Cairo to Suez, 93 miles, across the desert, in about 5 days. At Suez would be a station for steamers, and the Red Sea would be traversed in 3 runs or more. If necessary, stations for coals might be established at Cosseir, Judda, Mocha, and finally at Socatra—an island immediately beyond the mouth of the Red Sea, in the Indian Ocean: the run from Suez to Cosseir would be 300 miles—somewhat more than twice the distance from Liverpool to Dublin. From Cosseir to Judda, 450 miles; from Judda to Mocha, 517 miles; and from Mocha to Socatra, 632 miles. It is evident that all this would, without difficulty, in the most unfavourable weather, fall within the present powers of steam navigation. If the terminus of the passage be Bombay, the run from Socatra to Bombay will be 1200 miles, which would be, upon an average of weather about 8 days' steaming. The whole passage from Alexandria to Bombay, allowing 3 days for delay between Suez and Bombay, would be 26 days: the time from Bombay to Malta would therefore be about 33 days: and adding 14 days to this for the transit from Malta to England, we should have a total of 47 days from London to Bombay, or about 7 weeks.

If the terminus proposed were Calcutta, the course from Socatra would be 1250 miles south-east to the Maldives, where a station for coals would be established. This distance would be equal to that from Socatra to Bombay. From the Maldives, a run of 400 miles would reach the southern point of Ceylon, called the Point de Galle, which is the best harbour (Bombay excepted) in British India: from the Point de Galle, a run of 600 miles will reach Madras; and from Madras to Calcutta would be a run of about 600 miles. The voyage from London to Calcutta would be performed in about 60 days.

At a certain season of the year there exists a powerful physical opponent to the transit from India to Suez: from the middle of June until the end of September, the south-west monsoon blows with unabated force across the Indian Ocean, and more particularly between Socatra and Bombay. This wind is so violent as to leave it barely possible for the most powerful steam packet to make head against it, and the voyage could not be accomplished without serious wear and tear upon the vessels during these months—if indeed it would be practicable at all for any continuance in that season. The attention of Parliament has therefore been directed to another line of communication, not liable to this difficulty: it is proposed to establish a line of steamers from Bombay through the Persian Gulf to the Euphrates. The run from Bombay to a place called Muscat, on the southern shore of the Gulf, would be 840 miles in a north-west direction, and therefore not opposed to the south-west monsoon. From Muscat to Bassidore, a point upon the northern coast of the strait at the mouth of the Persian Gulf, would be a run of 255 miles; from Bassidore to Bushire, another point on the eastern coast of the Persian Gulf, would be a run of 300 miles; and from Bushire to the mouth of the Euphrates, would be 120 miles. It is evident that the longest of these runs would offer no more difficulty than the passage from Malta to Alexandria. From Bussora near the mouth of the Euphrates, to Bir, a town upon its left bank near Aleppo, would be 1143 miles, throughout which there are no physical obstacles to the river navigation which may not be overcome. Some difficulties arise from the wild and savage character of the tribes who occupy its banks. It is, however, thought that by proper measures, and securing the co-operation of the Pacha of Egypt, any serious obstruction from this cause may be removed. From Bir, by Aleppo, to Scanderoon, a port upon the Mediterranean, opposite Cyprus, is a land journey, said to be attended with some difficulty, but not of great length; and from Scandaroon to Malta is about the same distance as between the latter place and Alexandria. It is calculated that the time from London to Bombay by the Euphrates—supposing the passage to be successfully established—would be a few days shorter than by Egypt and the Red Sea.

Whichever of these courses may be adopted, it is clear that the difficulties, so far as the powers of the steam engine are concerned, lie in the one case between Socatra and Bombay, or between Socatra and the Maldives, and in the other case between Bombay and Muscat. Even the run from Malta to Alexandria or Scandaroon is liable to objection, from the liability of the boiler to deposite and incrustation, unless some effectual method be taken to remove this source of injury. If, however, the contrivance of Mr. Hall, or of Mr. Howard, or any other expedient for a like object, be successful, the difficulty will then be limited to the necessary supply of coals for so long a voyage. This, however, has already been encountered and overcome on four several voyages by the HUGH LINDSAY steamer from Bombay to Suez: that vessel encountered a still longer run on these several trips, by going, not to Socatra, but to Aden, a point on the coast of Arabia near the Straits of Babel Mandeb, being a run of 1641 miles, which she performed in 10 days and 19 hours. The entire distance from Bombay to Suez was in one case performed in 16 days and 16 hours; and under the most unfavourable circumstances, in 23 days. The average was 21 days for each trip.

(125.) Another projected line of steam communication, which offers circumstances of equal interest to the people of these countries and the United States, is that which is proposed to be established between London and New York. On the completion of the London and Liverpool railroad, Dublin will be connected with London by a continuous line of steam transport. It is proposed to continue this line by a railroad from Dublin to some point on the western coast of Ireland; among others, the harbour of VALENTIA has been mentioned. The nearest point of the western continent is

St. John's, Newfoundland, the distance of which from Valentia is 1900 miles; the distance from St. John's to New York is about 1200 miles, Halifax (Nova Scotia) being a convenient intermediate station. The distance from Valentia to St. John's comes very near the point which we have already assigned as the probable present limit of steam navigation. The Atlantic Ocean also offers a formidable opponent in the westerly winds which almost constantly prevail in it. These winds are, in fact, the reaction of the trades, which blow near the equator in a contrary direction, and are produced by those portions of the equatorial atmosphere which, rushing down the northern latitudes, carry with them the velocity from west to east proper to the equator. Besides this difficulty, St. John's and Halifax are both inaccessible, by reason of the climate, during certain months of the year. Should these causes prevent this project from being realized, another course may be adopted. We may proceed from the southern point of Ireland or England to the Azores, a distance of about 1800 miles: from the Azores to New York would be a distance of about 2000 miles, or from the Azores to St. John's would be 1600 miles.*

(*k*) While the inhabitants of Great Britain are discussing the project of the communication with New York, by means

* A treatise on the steam engine is not the place to enter into discussion on the causes of the several constant, periodic, and prevailing winds, otherwise we should feel it our duty to correct the opinion adopted by Dr. Lardner from the older authorities, in relation to the course of the westerly winds. These winds are, in the South Atlantic, and in both South and North Pacific, constant winds. In the North Atlantic, between the latitudes of 35° and 45°, and, therefore, in the track of the vessels which navigate between the United States and Great Britain, they are the most frequent prevailing winds, except in the months of April and October. They are certainly not the reaction of the trade winds, which is in a well-known zone, to the south of the region in which these westerly winds prevail, under the name of the Horse latitudes of our navigations, and the Grassy sea of the Spaniards. Those who wish to study the true theory of these winds will find it in Daniell's learned and ingenious work, "On Atmospheric Phenomena," or in the analysis of that work in the American Quarterly Review.

of the stations described by Dr. Lardner, those of the United States appear to be seriously occupied in carrying into effect a direct communication from New York to Liverpool. At the speed which has been given to the American steamboats, this presents no greater difficulties than the voyage from the Azores to New York, would, to one having the speed of no more than $7\frac{1}{4}$ miles per hour. As this attempt is beyond the limit of individual enterprise, there is, at the present moment, an application before the Legislature of the State of New York for a charter to carry this project into effect. It will be difficult to estimate the results of this enterprise, which will bring the old and new world within 12 or 15 days' voyage of each other.—A. E.*

* Dr. Lardner appears purposely to have omitted any detail of the history of Steam Navigation. It would be an invidious task on the part of a mere editor to attempt to supply what he has thought proper to avoid. We therefore merely refer to this subject for the purpose of expressing the hope, that this silence is an earnest that the writers of Europe are about to abandon the claims they have set up for their countrymen to the merit of introducing the successful practice of steam navigation, and that the respective services of Fitch, Evans, Fulton, and the elder Stevens will soon be universally acknowledged.—A. E.

# CHAPTER XII.

## GENERAL ECONOMY OF STEAM-POWER.

Mechanical Efficacy of Steam—Proportional to the Quantity of Water evaporated, and to the Fuel consumed—Independent of the Pressure.—Its mechanical Efficacy by Condensation alone—By Condensation and Expansion combined—By direct Pressure and Expansion—By direct Pressure and Condensation—By direct Pressure, Condensation, and Expansion.—The Power of Engines.—The Duty of Engines.—Meaning of Horse-power.—To compute the Power of an Engine.—Of the Power of Boilers.—The Structure of the Grate Bars.—Quantity of Water and Steam Room.—Fire Surface and flue Surface.—Dimensions of Steam Pipes.—Velocity of Piston.—Economy of Fuel.—Cornish Duty Reports.

(130.) HAVING explained in the preceding chapters the most important circumstances connected with the principal varieties of steam engines, it remains now to explain some matters of detail connected with the power, efficiency, and economy of these machines, which, though perhaps less striking and attractive than the subjects which have hitherto engaged us, are still not undeserving of attention.

It has been shown in the first chapter, that water exposed to the ordinary atmospheric pressure (the amount of which may be expressed by a column of 30 inches of mercury) will pass from the liquid into the vaporous state when it arrives at the temperature of 212°; and the vapour thus produced from it will have an elastic force equal to that of the atmosphere. If the water, however, to which heat is applied, be submitted to a greater or less pressure than that of the atmosphere, it will boil at a greater or less temperature, and will always produce steam of an elastic force equal to the pressure under which it boils. Now it is a fact as remarkable as it is important, that to convert a given weight of water into vapour will require the same quantity of heat, under whatever pressure, and at whatever temperature the water may boil. Let us suppose a tube, the base of which is equal to a square foot, in which a piston fits air-tight and steam-tight. Immediately under the piston, let a cubic inch of water be placed, which will be spread in a thin layer over the bottom of the tube. Let the piston be counterbalanced by a weight (acting over a pulley) which will be equivalent to the weight of the piston, so that it shall be free to ascend by the application of any pressure below it. Now let the flame of a lamp be applied at the bottom of the tube: the water under the piston being affected by no pressure from above, except that of the atmosphere acting upon the piston, will boil at the temperature of 212°, and by the continued application of the lamp it will at length be converted into steam. The steam into which the cubic inch of water is converted will expand into the magnitude of a cubic foot, exerting an elastic force equal to the atmospheric pressure; consequently the piston will be raised one foot above its first position in the tube, and the cubic foot beneath it will be completely filled with steam. Let us suppose, that to produce this effect required the lamp to be applied to the tube for the space of fifteen minutes.

The water being again supposed at its original temperature, and the piston in its first position, let a weight be placed upon the piston equal to the pressure of the atmosphere, so that the water beneath the piston will be pressed down by double the atmospheric pressure. If the lamp be once more applied, the water will, as before, be converted into vapour; but the piston will now be raised to the height of only six inches* from the bottom, the steam expanding into only half its former bulk. The temperature at which

* Strictly speaking, the height to which the piston would be raised would not diminish in so great a proportion as the pressure is increased, because the increase of pressure being necessarily accompanied by an increase of temperature, a corresponding expansion would be produced. Therefore there will be a slight increase in the total mechanical effect of the steam. The difference, however, is not very important in practice, and it is usual to consider the density of steam as proportional to the pressure.

the water would commence to be converted into vapour, instead of being 212°, would be 250°; but the time elapsed between the moment of the first application of the lamp and the complete conversion of the water into steam, will still be fifteen minutes.

Again, if the piston be loaded with a weight equal to double the atmospheric pressure, the water will be pressed down by the force of three atmospheres. If the lamp be applied as before, the water would be converted into steam in the same time; but the piston will now be raised only four inches above its first position, and the steam will consequently be three times as dense as when the piston was pressed down only by the atmosphere.

From these and similar experiments we infer:—

*First*, That the elastic pressure of steam is equal to the mechanical pressure under which the water producing the steam has been boiled.

*Secondly*, That the bulk which steam fills is diminished in the same proportion as the pressure of the steam is increased; or, in other words, that the density of steam is always in the same proportion as its pressure.

*Thirdly*, That the same quantity of heat is sufficient to convert the same weight of water into steam, whatever be the pressure under which the water is boiled, or whatever be the density and pressure of the steam produced.

*Fourthly*, That the same quantity of water being, converted into steam, produces the same mechanical effect, whatever be the pressure or the density of the steam. Thus, in the first case, the weight of one atmosphere was raised a foot high; in the second case, the weight of two atmospheres was raised through half a foot; and, in the third case, the weight of three atmospheres was raised through the third of a foot; the weight raised being in every case increased in the same proportion as the height through which it is elevated is diminished. Every increase of the weight is, therefore, compensated by a proportionate diminution of the height through which it is raised, and the mechanical effect is consequently the same.

*Fifthly*, That the same quantity of heat or fuel is necessary and sufficient to produce the same mechanical effect, whatever be the pressure of the steam which it produces.

If steam be used to raise a piston against the atmospheric pressure only, although a definite physical force will be exerted by it, and a mechanical effect produced, yet under such circumstances it will exert no directly useful efficiency; but after the piston has been raised, and the tube beneath it filled with steam balancing the atmosphere above it, a useful effect to the same amount may be obtained by cooling the tube, and thereby reconverting the steam into water. The piston will thus be urged downward by the unresisted force of the atmosphere, and any chain or rod attached to it will be drawn downward with a corresponding force. If the area of the piston be, as already supposed, equal to the magnitude of one square foot, the atmospheric pressure upon it, being 15 pounds for each square inch, will amount to 144 times 15 pounds, or 2160 pounds. By drawing down a chain or rope acting over a pulley, the piston would in its descent (omitting the consideration of friction, &c.) raise a weight of 2160 pounds a foot high. Since 2160 pounds are nearly equal to one ton, it may, for the sake of round numbers be stated thus:—

"*A cubic inch of water, being converted into steam, will, by the condensation of that steam, raise a ton weight a foot high.*" Such is the way in which the force of steam is rendered practically available in the atmospheric engine.

(131.) The method by which steam is used in the single-acting steam engine of Watt is, in all respects, similar, except that the piston, instead of being urged downward by the force of the atmosphere, is pressed by steam of a force equal to the atmospheric pressure. It is evident, however, that this does not alter the mechanical result.

We have stated that a considerable increase of power, from a given quantity of steam, was produced by cutting off the steam after the piston had made a part of its descent, and allowing the remainder of the descent to be produced by the expansive force of the steam already admitted. We shall now more fully explain the principle on which this increase of power depends.

Let A B, (fig. 69,) as before, represent a tube, the bottom of which is equal to a square foot, and let P be a piston in it, resting upon a cubic inch of water spread over the bottom; and let W be an empty vessel, the weight of which exactly counterpoises the piston. By the application of the lamp, the water will be converted into steam of the atmospheric pressure, and the piston will be raised from P to P′, through the height of one foot, the space in the tube beneath it being filled with steam, and the vessel W will have descended through one foot. Let half a ton of water be now poured into the vessel W; its weight will draw the piston P′ upward, so that the steam below it will expand into a larger space. When the piston P′ was only balanced by the empty vessel W, it was pressed downward by the whole weight of the atmosphere above, which amounts to about one ton: now, however, half of this pressure is balanced by the half ton of water poured into the vessel W; consequently the effective downward pressure on the piston P′ will be only half a ton, or half its former amount. The piston will therefore rise, until the pressure of the steam below it is diminished to the same extent. By what has been already explained, this will take place when the steam is allowed to expand into double its former bulk; consequently, when the piston has risen to P″, one foot higher, or two feet from the bottom of the tube, the steam will then exactly balance the downward pressure on the piston, and the latter will remain stationary; the vessel W, with the half ton of water it contains, will have descended one foot lower, or two feet below its first position. Let the steam now be cooled and reconverted into water, and at the same time let another half ton of water be supplied to the vessel W; the pressure below the piston being entirely removed, the atmospheric pressure will act above it with undiminished force; and this force, amounting to one ton, will draw up the vessel W, with its contents. When the piston descends, as it will do, to the bottom of the tube, the ton of water contained in the vessel W will be raised through two perpendicular feet.*

Now, in this process it will be observed that the quantity of steam consumed is not more than in the former case, viz. the vapour produced by boiling one cubic inch of water. Let us consider, however, the mechanical effect which has resulted from it; half a ton of water has been allowed to descend through one foot, while a ton has been raised through two feet: deducting the force lost by the descent of half a ton through one foot from the force obtained by the ascent of one ton through the two feet, we obtain for the whole mechanical effect one ton and a half raised through one foot; for it is evident that half a ton has been raised from the lowest point to which the vessel W descended one foot above that point, and one ton has been raised through the other foot, which is equivalent to one ton and a half through one foot.

Comparing this with the effect produced in the first case, where the steam was condensed without causing its expansion, it will be evident that there is an increase of 50 per cent. upon the whole mechanical effect produced.

But this is not the limit of the increase of power by expansion. Instead of condensing the steam when the piston had arrived at P'', let a further quantity of water amounting to one sixth of a ton be poured into the vessel W, in addition to the half ton which it previously contained; the

* Strictly speaking, the quantity of water supposed in these cases to be placed in the vessel W would just *balance* the atmospheric pressure. A slight preponderance must therefore be given to the piston, to produce the motion.

effective pressure on the piston P'', being only half a ton, will be overbalanced by the preponderating weight in the vessel W, and the piston will consequently ascend. It will become stationary when the steam by expansion loses a quantity of force equal to the additional weight which the vessel W has received: now, that vessel, having successively received a half and a sixth of a ton, will contain two-thirds of a ton; consequently the effective downward pressure on the piston will be only a third of a ton, and the steam to balance this must expand into three times the space it occupied when equal to the atmospheric pressure. It must therefore ascend to P''', three feet above the bottom of the tube. If the steam in the tube be now condensed, and at the same time one-third of a ton of water be supplied to the vessel W, so as to make its total contents amount to one ton, the piston will descend, being urged downward by the unresisted atmospheric pressure, and the ton of water contained in the vessel W will be raised through three perpendicular feet.

In this case, as in the former, the total quantity of steam consumed is that of one cubic inch of water; but the mechanical effect it produces is still further increased. To calculate its amount, we must consider that half a ton of water has fallen through two feet, which is equivalent to a ton falling through one foot, besides which the sixth part of a ton has fallen through one foot. The total loss, therefore, by the fall of water has been one ton and one sixth through one foot, while the force gained by the ascent of water has been one ton raised through three feet, which is equivalent to three tons through one foot. If, then, from three tons we deduct one and one-sixth, the remainder will be one ton and five-sixths raised through one foot; this effect being above 80 per cent. more than that which is produced in the first case, where the steam was not allowed to expand.

To carry the inquiry one step further: Let us suppose that, upon the arrival of the piston at P''', a further addition of water to the amount of one-twelfth of a ton be added to it: this, with the water it already contained, would make the total contents three-fourths of a ton; consequently, the effective pressure upon the piston would now be reduced to one-fourth of the atmospheric pressure. The atmospheric steam would balance this when expanded into four times its original volume: consequently, the piston would come to a state of rest at P'''', four feet above the bottom of the tube, and the vessel W would consequently have descended through four perpendicular feet. If the steam in the tube be now condensed as in the former cases, and at the same time a quarter of a ton of water be added to the vessel W, the piston will descend to the bottom of the tube, and the ton of water in the vessel W will be raised through four perpendicular feet. To estimate the mechanical effect thus produced, we have, as before, to deduct the total force lost by the fall of water from the force gained by its elevation: the water has fallen in three distinct portions: first, half a ton has fallen through three perpendicular feet, which is equivalent to one ton and a half through one foot; secondly, one-sixth of a ton has fallen through two perpendicular feet, which is equivalent to one-third of a ton through one foot; and thirdly, one-twelfth of a ton has fallen through one foot: these added together will be equivalent to one ton and eleven-twelfths through one foot. One ton has been raised through four feet, which is equivalent to four tons through one foot: deducting from this the force lost by the descent, the surplus gained will be two tons and one-twelfth through one foot, being about 108 per cent. more than the force resulting from the condensation of steam without expansion.

To the increase of mechanical effect to be produced in this way, there is no theoretical limit. According to the manner in which we have here explained it, to produce the greatest possible effect by a given extent of expansion, it would be necessary to supply the water or other counterpoise to the vessel W, not in separate masses, as we have here supposed, but continuously, so as to produce a regular motion of the piston upward.

Such is the principle on which the advantages of the expansive engine of Watt and Hornblower depend, explained so far as it can be without the aid of the language and reasoning of analysis.*

(132.) We have here, however, only considered the mechanical effect produced by the condensation of steam. Let us now examine its direct action.

Let the piston P be supposed to be connected by a rod with a load or resistance which it is intended to raise, and let the load placed upon it be supposed to amount to one ton, the total pressure on the piston will then be two tons; one due to the atmospheric pressure, and the other to the amount of the load. Upon applying heat to the water, steam will be produced; and when the water has been completely evaporated, the piston will rise to the height of six inches from the bottom of the tube. The total mechanical effect thus produced will be one ton weight raised through six perpendicular inches, which is equivalent to half a ton raised through one foot.

Again, let the load upon the piston be two tons; this will produce a total pressure upon the water below it amounting to three tons, including the atmospheric pressure. The water, when converted into vapour under this pressure, will raise the piston and its load through four perpendicular inches: the useful mechanical effect will then be two tons raised through the third of a foot, which is equivalent to two-thirds of a ton raised one foot. In the same manner, if the piston were loaded with three tons, the mechanical effect would be equivalent to three-fourths of a ton raised through one foot, and so on.

* A strict investigation of this important property, as well as of the other consequences of the quality of expansion, would require more abstruse mathematical processes than would be consistent with the nature of this work.

It appears therefore from this reasoning, that when the direct force of steam of greater pressure than the atmosphere is used without condensation, the total mechanical effect is always less than that produced by the condensation of atmospheric steam without expansion; but that the greater the pressure under which the steam is produced, the less will be the difference between these effects. In general, the proportion of the mechanical effect of high-pressure steam to the effect produced by the condensation of atmospheric steam, will be as the number of atmospheres expressing the pressure of the steam to the same number increased by one. Thus, if steam be produced under the pressure of six atmospheres, the proportion of its effect to that of the condensation of atmospheric steam will be as six to seven.

(133.) Another method of applying the power of steam mechanically is, to combine its direct action with condensation, but without expansion.

The piston being, as before, loaded with one ton, the evaporation of the water will raise it through six perpendicular inches, and the result so far will be equivalent to a ton raised half a foot; but if the piston-rod be supposed also to act by a chain or cord over a wheel, so as to pull a weight up, the steam which has just raised the ton weight through six inches, may be condensed, and the piston will descend with a force of one ton into the vacuum thus produced, and another ton may be thus raised through half a foot. The total mechanical power thus yielded by the steam, adding to its direct action its effect by condensation, will then be one ton raised through one foot, being an effect exactly equal to that obtained by the condensation of atmospheric steam.

If the piston be loaded with two tons, its direct action will, as we have shown, raise these two tons through four inches, which is equivalent to two-thirds of a ton raised a foot. By condensing this steam a ton weight may be raised in the same manner, by the descent of the piston through a third of a foot, which is equivalent to the third of a ton raised through one foot.

By pursuing like reasoning, it will appear that, if the direct force of high-pressure steam be combined with the indirect force produced by its condensation, the total mechanical effect will be precisely equal to the mechanical effect by the mere condensation of atmospheric steam.

(134.) In applying the principle of expansion to the direct action of high-pressure steam, advantages are gained analogous to those already explained with reference to the method of condensation.

Let the piston be supposed to be loaded with three tons; the evaporation of the water beneath it will raise this weight, including the atmospheric pressure, through three perpendicular inches. Let one ton be now removed, and the remaining two tons will be raised, by the expansion of the steam, through another perpendicular inch. Let the second ton be now removed, and the piston loaded with the remaining ton will rise, by the expansion of the steam, to the height of six inches from the bottom. These consequences follow immediately from the principle that steam will expand in proportion as the pressure upon it is diminished, observing that in this case the atmospheric pressure, amounting to one ton, must always be added to the load. In this process three separate effects are produced: one ton is raised through three inches, which is equivalent to a quarter of a ton raised through one foot; another ton is raised through four inches, which is equivalent to a third of a ton through a foot, and the third ton is raised through six inches, which is equivalent to half a ton raised through a foot. The total of these effects amounts to one and one-twelfth of a ton raised through one foot, while the same load, raised by the high-pressure steam without expansion, would be equivalent to only half a ton raised through one foot.

Again, let the load placed upon the piston be five tons the evaporation of the water will raise this through the sixth part of a foot; if one ton be now removed, the other four tons will be raised to a height above the bottom of the tube equal to a fifth part of a foot; another ton being removed, the remaining three will be raised to a height from the bottom equal to a fourth of a foot; and so on, the last ton being raised through half a foot. To estimate the total mechanical effect thus produced, we are to consider that the several tons raised from their first position are raised through the sixth, fifth, fourth, third, and half of a perpendicular foot, giving a total effect equal to the sixth, fifth, fourth, third, and half of a ton severally raised through one foot; these, therefore, added together, will give a total of nineteen-twentieths of a ton raised through one foot.

In general, the expansive force applied to the direct action of high-pressure steam, therefore, will increase its effect according to the same law, and subject to the same principles as were shown with respect to the method of condensation accompanied with expansion.

The expansive action of high-pressure steam may be accompanied with condensation, so as considerably to increase the mechanical effect produced; for, after the weights with which the piston is loaded have been successively raised to the extent permitted by the elastic force of the steam, and are removed from the piston, the steam will expand until it balances the atmospheric pressure. It may afterwards be made further to expand, by adding weights to the counterpoise w in the manner already explained; and, the steam being subsequently condensed, all the effects will be produced upon the descent of the piston which we have before noticed. It is evident that by this means the mechanical effect admits of very considerable increase.

(135.) We have hitherto considered the piston to be resisted by the atmospheric pressure above it; but, as is shown in the preceding chapters, in the modern steam engines the atmosphere is expelled from the interior of the machine by allowing the steam to pass freely through all its cavities in the first instance, and to escape at some convenient aperture, which, opening outwards, will effectually prevent the subsequent readmission of air. The piston-rod and other parts which pass from the external atmosphere to the interior of the machine, are likewise so constructed and so supplied with oil or other lubricating matter, that neither the escape of steam nor the entrance of air is permitted. We are therefore now to consider the effect of the action of steam against the piston P, when subjected to a resistance which may be less in amount, to any extent, than the atmospheric pressure.

In such machines the steam always acts both directly by its power, and indirectly by its condensation. In calculating its effects, excluding friction, &c., we have therefore only to estimate its total force upon the piston, and to deduct the force of the uncondensed vapour which will resist the motion of the piston.

Supposing, then, the total force exerted upon the piston, after deducting the resistance from the uncondensed vapour, to be one ton, and the length of the cylinder to be one foot, each motion of the piston from end to end of the cylinder will produce a mechanical force equivalent to a ton weight raised one foot high. If in this case the magnitude of the piston be equivalent to one square foot, the pressure of the steam will be equal to that of the atmosphere, and the quantity of water in the form of steam which the cylinder will contain will be a cubic inch, while the quantity of steam in it will be a cubic foot. In proportion as the area of the piston is enlarged, the pressure of the steam will be diminished, if the moving force is required to remain the same; but with every diminution of pressure the density of the steam will be diminished in the same proportion, and the cylinder will still contain the same quantity of water in the form of vapour. In this way steam may be used, as a mechanical agent, with a pressure to almost any extent less than that of the atmosphere, and at temperatures considerably lower than 212°. To obtain the same mechanical force, it is only necessary to enlarge the piston in the same proportion as the pressure of the steam is diminished.

By a due attention to this circumstance, the expansive power of steam, both in its direct action and by condensation, may be used with very much increased advantage; and such is the principle on which the benefits derived from Woolf's contrivances depend. If steam of a high pressure, say of three or four atmospheres, be admitted to the piston, and allowed to impel it through a very small portion of the descent, it may then be cut off, and its expansion may be allowed to act upon the piston until the pressure of the steam is diminished considerably below the atmospheric pressure; the steam may then be condensed and a vacuum produced, and the process repeated.

In the double-acting engines, commonly used in manufactures and in navigation, and still more in the high-pressure engines used for locomotion, the advantageous application of the principle of expansion appears to have been hitherto attended with difficulties; for, notwithstanding the benefits which unquestionably attend it in the economy of fuel, it has not been generally resorted to. To derive from this principle full advantage, it would be necessary that the varying power of the expanding steam should encounter a corresponding, or a nearly corresponding, variation in the resistance: this requisite may be attained, in engines applied to the purpose of raising water, by many obvious expedients; but when they have, as in manufactures, to encounter a nearly uniform resistance, or, in navigation and locomotion, a very irregular resistance, the due application of expansion is difficult, if indeed it be practicable.

We have seen that the mechanical effect produced by steam when the principle of expansion is not used, is always proportional to the quantity of water contained in the steam, and is likewise in the same proportion so long as a given degree of expansion is used. It is apparent, therefore, that the mechanical power which is or ought to be exerted by an engine is in the direct proportion of the quantity of water evaporated. It has also been shown that the quantity of water evaporated, whatever be the pressure of the steam, will be in the direct proportion of the quantity of heat received from the fuel, and therefore in the direct proportion of the quantity of fuel itself, so long as the same proportion of its heat is imparted to the water.

(136.) The POWER of an engine is a term which has been used to express the rate at which it is able to raise a given load, or overcome a given resistance. The DUTY of an engine is another term, which has been adopted to express the load which may be raised a given perpendicular height, by the combustion of a given quantity of fuel.

When steam engines were first introduced, they were commonly applied to work pumps or mills which had been previously wrought by horses. It was, therefore, convenient, and indeed necessary, in the first instance, to be able to express the performance of these machines by reference to the effects of animal power, to which manufacturers, miners, and others had been long accustomed. When an engine, therefore, was capable of performing the same work in a given time, as any given number of horses of average strength usually performed, it was said to be an engine of so many *horses' power*. This term was long used with much vagueness and uncertainty : at length, as the use of steam engines became more extended, it was apparent that confusion and inconvenience would ensue, if some fixed and definite meaning were not assigned to it, so that the engineers and others should clearly understand each other in expressing the powers of these machines. The term *horse-power* had so long been in use, that it was obviously convenient to retain it. It was only necessary to agree upon some standard by which it might be defined. The performance of a horse of average strength, working for eight hours a day, was, therefore, selected as a standard or unit of steam engine power. Smeaton estimated the amount of mechanical effect which the animal could produce at 22,916 pounds, raised one foot per minute; Desaguiliers makes it 27,500 pounds, raised through the same height. Messrs. Bolton and Watt caused experiments to be made with the strong horses used in the breweries in London; and from the result of these they assigned 33,000 pounds raised one foot per minute, as the value of a horse's power: this is, accordingly, the estimate now generally adopted; and, when an engine is said to be of so many horses' power, it is meant that, when in good working order and properly managed, it is capable of overcoming a resistance equivalent to so many times 33,000 pounds raised one foot per minute. Thus, an engine of ten-horse power would be capable of raising 330,000 pounds one foot per minute.

As the same quantity of water converted into steam will always produce the same mechanical effect, whatever be the density of the steam produced from it, and at whatever rate the evaporation may proceed, it is evident that the *power* of a steam engine will depend on two circumstances: first, the rate at which the boiler with its appendages is capable of evaporating water; and, secondly, the rate at which the engine is capable of consuming the steam by its work. We shall consider these two circumstances separately.

The rate at which the boiler produces steam will depend upon the rate at which heat can be transmitted from the fire to the water which it contains. Now this heat is transmitted in two ways: either by the direct action of the fire radiating heat against the surface of the boiler, or by the flame and heated air which escapes from the fire passing through the flues, as already explained. The surface of the boiler exposed to the direct radiation of the fire is technically called *fire surface;* and that which takes heat from the flame and air, on its way to the chimney, is called *flue surface.* Of these the most efficient in the generation of steam is the former. In stationary boilers, used for condensing engines, where magnitude and weight are matters of little importance, it has been found that the greatest effect has been produced in general by allowing four and a half square feet of fire surface, and four and a half square feet of flue surface, for every horse-power. By means of this quantity of fire and flue surface, a cubic foot of water per hour may be evaporated.

It has already been shown that the total power exerted by a cubic inch of water, converted into steam, will be equivalent to 2160 pounds raised one foot. A cubic foot of water consists of 1728 cubic inches, and the power produced by its evaporation will therefore be found by multiplying 216 by 1728; the product, 3,732,480, expresses the number of pounds' weight which the evaporation of a cubic foot of water would raise one foot high, supposing that its entire mechanical force were rendered available: but to suppose this in practice, would be to suppose the machine, through the medium of which it is worked, moved without any power being expended upon its own parts. It would be, in fact, supposing all its moving parts to be free from friction and other causes of resistance. To form a practical estimate, then, of the real quantity of available mechanical power obtained from the evaporation of a given quantity of water, it will be necessary to inquire what quantity of this power is intercepted by the engine through which it is transmitted. In different forms of steam engine—indeed, we may say in every individual steam engine—the amount thus lost is different; nevertheless, an approximate estimate may be obtained, sufficiently exact to form the basis of a general conclusion.

Let us consider, then, severally, the means by which mechanical power is intercepted by the engine.

*First*, The steam must flow from the boiler into the cylinder to work the piston; it passes necessarily through pipes more or less contracted, and is, therefore, subject to friction as well as cooling in its passage.

*Second*, Force is lost by the radiation of heat from the cylinder and its appendages.

*Third*, The friction of the piston in the cylinder must be overcome.

*Fourth*, Loss of steam takes place by leakage.

*Fifth*, Force is expended in expelling the steam after having worked the piston.

*Sixth*, Force is required to open and close the several valves, to pump up the water for condensation, and to overcome the friction of the various axles.

*Seventh*, Force is expended upon working the air-pump.

In engines which do not condense the steam, and which, therefore, work with steam of high pressure, some of these sources of waste are absent, but others are of increased amount. If we suppose the total effective force of the water evaporated per hour in the boiler to be expressed by 1000, it is calculated that the waste in a high-pressure engine will be expressed by the number 392 ; or, in other words, taking the whole undiminished force obtained by evaporation as expressed by 10, very nearly 4 of these parts will be consumed in moving the engine, and the other 6 only will be available.

In a single-acting engine which condenses the steam, taking, as before, 1000 to express the total mechanical power of the water evaporated in the boiler, 402 will express the part of this consumed in moving the engine, and 598, therefore, will express the portion of the power practically available ; or, taking round numbers, we shall have the same result as in the non-condensing engine, viz. the whole force of the water evaporated being expressed by 10, 4 will express the waste, and 6 the available part.

In a double-acting engine the available part of the power bears a somewhat greater proportion to the whole. Taking, as before, 1000 to express the whole force of the water evaporated, 368 will express the proportion of that force expended on the engine, and 632 the proportion which is available for work.

In general, then, taking round numbers, we may consider

that the mechanical force of four-tenths of the water evaporated in the boiler is intercepted by the engine, and the other six-tenths are available as a moving force. In this calculation, however, the resistance produced in the condensing engine by the uncondensed steam is not taken into account: the amount of this force will depend upon the temperature at which the water is maintained in the condenser. If this water be kept at the temperature of 120°, the vapour arising from it will have a pressure expressed by three inches seven-tenths of mercury; if we suppose the pressure of steam in the boiler to be measured by 37 inches of mercury, then the resistance from the uncondensed steam will amount to one-tenth of the whole power of the boiler; this, added to the four-tenths already accounted for, would show a waste amounting to half the whole power of the boiler, and consequently only half the water evaporated would be available as a moving power.

If the temperature of the condenser be kept down to 100°, then the pressure of uncondensed steam will be expressed by two inches of mercury, and the loss of power consequent upon it would amount to a proportionally less fraction of the whole power.

The following example will illustrate the method of estimating the effective power of an engine.

In a double-acting engine, in good working condition, the total power of steam in the boiler being expressed by 1000, the proportion intercepted by the engine, exclusive of the resistance of the uncondensed steam, will be 368, and the effective part 632. Now, suppose the pressure of steam in the boiler to be measured by a column of 35 inches of mercury; the thousandth part of this will be seven two-hundredths of an inch of mercury, and 632 of these parts will express the effective portion of the power. By multiplying seven two-hundredths by 632, we obtain 22 nearly. Now, suppose the temperature in the condenser is 1200, the pressure of steam corresponding to that temperature will be measured by $3\frac{7}{10}$ inches of mercury. Subtracting this from 22, there will remain $18\frac{3}{10}$ inches of mercury, as the effective moving force upon the piston; this will be equivalent to about 7 lbs. on each circular inch.

If the diameter of the piston then be 24 inches, its surface will consist of a number of circular inches expressed by the square of 24, or $24 \times 24 = 576$; and, as upon each of these circular inches there is an effective pressure of 7 lbs., we shall find the total pressure in pounds by multiplying 576 by 7, which gives 4032 lbs.

We shall find the space through which this force works per minute, by knowing the length of the cylinder and the number of strokes per minute. Suppose the length of the cylinder to be 5 feet, and the number of strokes per minute $21\frac{1}{2}$. In each stroke* the piston will, therefore, move through 10 feet, and in one minute it will move through 215 feet. The moving force, therefore, is 4032 lbs. moved through 215 feet per minute, which is equivalent to 215 times 4032 lbs., or 866,880 lbs., raised one foot per minute.

For every 33,000 lbs. contained in this, the engine has a horse-power. To find the horse-power, then, of the engine, we have only to divide 866,880 by 33,000; the quotient is 26 nearly, and, therefore, the engine is one of 26 horse-power.

Let it be required to determine the quantity of water which a boiler must evaporate per hour, for each horse-power of the engine which it works.

It has been already explained that one horse-power expresses 33,000 lbs. raised one foot high per minute, or, 1,980,000 lbs. raised one foot high per hour. The quantity of water necessary to produce this mechanical effect by evaporation, will be found by considering that a cubic inch of water, being evaporated, will produce a mechanical force

* By a stroke of the piston is meant its motion from one end of the cylinder and back again.

equivalent to 2160 lbs. raised a foot high. If we divide 1,980,000, therefore, by 2160, we shall find the number of cubic inches of water which must be evaporated per hour, in order to produce the mechanical effect expressed by one horse-power; the result of this division will be 9166, which is therefore the number of cubic inches of water per hour, whose evaporation is equivalent to one horse-power. But it has been shown that, for every 6 cubic inches of water evaporated in the boiler which are available as a moving power, there will be 4 cubic inches intercepted by the engine. To find, then, the quantity of waste corresponding to 916 cubic inches of water, it will be necessary to divide that number by 6, and to multiply the result by 4: this process will give 610 as the number of cubic inches of water wasted. The total quantity of water, therefore, which must be evaporated per hour, to produce the effect of one horse-power, will be found by adding 610 to 916, which gives 1526.

This result, however, being calculated upon a supposition of a degree of efficiency in the engines which is, perhaps, somewhat above their average state, it has been customary with engineers to allow a cubic foot of water per hour for each horse-power, a cubic foot being 1728 cubic inches, or above 11 per cent. more than the above estimate.

(137.) It has been stated, that to evaporate a cubic foot of water per hour requires 9 square feet of surface exposed to the action of the fire and heated air. This, therefore, is the quantity of surface necessary for each horse-power, and we shall find the total quantity of fire and flue surface necessary for a boiler of a given power, by multiplying the number of horses in the power by 9; the product will express, in square feet, the quantity of boiler surface which must be exposed to the fire, one-half of this being fire surface and the other half flue surface.

Since the supply of heat to the boiler must be proportionate to the quantity of fuel maintained in combustion, and the quantity of that fuel must depend on the extent of grate surface, it is clear that a determinate proportion must exist between the power of the boiler and the extent of grating in the fireplace. The quantity of oxygen which combines with the fuel varies with the quality of that fuel; for different kinds of coal it varies from two to three pounds for each pound of coal.

We shall take it an average of $2\frac{1}{2}$ pounds. Now $2\frac{1}{2}$ pounds of oxygen will measure 30 cubic feet; also 5 cubic feet of atmospheric air contain 1 cubic foot of oxygen; and consequently 150 cubic feet of atmospheric air will be necessary for the combustion of 1 pound of average coals. At least one-third of the air, which passes through a fire, escapes uncombined into the chimney. We must, therefore, allow 220 cubic feet of atmospheric air to pass through the grate bars for every pound of fuel which is consumed. Now since land boilers will consume 15 pounds, and marine boilers 10 pounds, per hour per horse-power, it follows that the spaces between the grate bars, and the extent of grate surface, must be sufficient to allow 3000 cubic feet of air per hour in land boilers, and 2000 cubic feet in marine boilers, to pass through them for each horse-power, or, what is the same, for each foot of water converted into steam per hour. The quantity of grate surface necessary for this does not seem to be ascertained with precision; but, perhaps, we may take as an approximate estimate for land boilers one square foot of grate surface per horse-power, and for marine boilers two-thirds of a square foot, the spaces between the grate bars being equal to their breadth.

It is evident that the capacity of a boiler for water and steam must have a determinate relation to the power of the engine it is intended to supply. For each horse-power of the engine, it has been shown that a cubic foot of water must pass from the boiler in the form of steam per hour. Now, it is evident that the steam could not be supplied of a uniform force, if the quantity of steam contained at any moment in the boiler were not considerably greater than the contents of the cylinder. For example, if the volume of steam in the boiler were precisely equal to the capacity of the cylinder, then one measure of the cylinder would for the moment cause the steam to expand into double its bulk and to lose half its force, supposing it to pass freely from the boiler to the cylinder. In the same manner, if the volume of steam contained in the boiler were twice the contents of the cylinder, the steam would for a moment lose a third of its force, and so on. It is clear, therefore, that the space allotted to steam in the boiler must be so many times greater than the magnitude of the cylinder, that the abstraction of a cylinder full of steam from it shall cause a very trifling diminution of its force.

In the same manner, we may perceive the necessity of maintaining a large proportion between the total quantity of water in the boiler, and the quantity supplied in the form of steam to the cylinder. If, for example, (taking as before an extreme case,) the quantity of water in the boiler were only equal to the quantity supplied in the form of steam to the cylinder in a minute, it would be necessary that the contents of the boiler should be replaced by cold water once in each minute: and, under such circumstances, it is evident that the action of the heat upon the water would be quite unmanageable. But, independent of this, the quantity of water must be sufficient to fill the boiler above the point at which the flue surface terminates, otherwise the heat of the fuel would act upon the part of the boiler containing steam and not water; and, steam receiving heat sluggishly, the metal of the boiler would be gradually destroyed by undue temperature.

The total quantity of space for water and steam in boilers is subject to considerable variation in proportion to their power. Small boilers require a greater proportion of steam and water room, or a greater capacity of boiler, in proportion, than large ones; and the same applies to their fire surface and flue surface.

The general experience of engineers has led to the conclu

sion, that a low-pressure boiler of the common kind requires ten cubic feet of water room, and ten cubic feet of steam room in the boiler, for every cubic foot which the engine consumes per hour, or, what is the same, for each horse-power of the engine. Thus, an engine of ten horse-power, according to this rule, would require a boiler having the capacity of 200 cubic feet which should be constantly kept half filled with water. There are, however, different estimates of this. Some engineers hold that a boiler should have twenty-five cubic feet of capacity for each horse-power of the engine, while others reduce the steam so low as eight cubic feet.

In a table of the capacities of boilers of different powers, and the feed of water necessary to be maintained in them, Mr. Tredgold assigns to a boiler of five horse-power fourteen cubic feet of water per horse-power; for one of ten horse-power, twelve and a half cubic feet; and, for one of forty horse, eleven cubic feet.

For engines of greater power it is generally found advantageous to have two or more boilers of small power, instead of one of large power. This method is almost invariably adopted on board steamboats, and has the advantage of securing the continuance of the working of the engine, in case of one of the boilers being deranged. It is also found convenient to keep an excess of power in the boilers, above the wants of the engine. Thus, an engine of sixty horse-power may be advantageously supplied with two forty horse boilers, and an engine of eighty horse-power with two fifty horse boilers, and so on.

(138.) The pressure of steam in the cylinder of an engine is always less than the pressure of steam in the boiler, owing to the obstructions which it encounters in its passage through the steam pipes and valves. The difference between these pressures will depend upon the form and magnitude of the passages: the straighter and wider they are, the less the difference will be; if they are contracted and subject to bends, especially to angular inflections, the steam will be considerably diminished in its pressure before it reaches the cylinder. The throttle valve placed in the steam pipe may also be so managed as to diminish the pressure of steam in the cylinder to any extent: this effect, which is well understood by practical engineers, is called *wire-drawing* the steam. By such means it is evidently possible for the steam in the boiler to have any degree of high pressure while the engine is worked at any degree of low pressure. Since, however, the pressure of the steam in the cylinder is a material element in the performance of the engine, the magnitude, position, and shape of the steam pipe and of the valves are a matter of considerable practical importance. But theory furnishes us with little more than very general principles to guide us. One practical rule which has been adopted is, to make the diameter of the steam pipe about one-fifth of that of the cylinder: by this means the area of the transverse action of the pipe will be one twenty-fifth part of the superficial magnitude of the piston; and, since the same quantity of steam per minute must flow through this pipe as through the cylinder, it follows that the velocity of the steam, in passing through the steam pipe, will be twenty-five times the velocity of the piston.

(139.) Another rule which has been adopted is, to allow a square inch of magnitude, in the section of the steam pipe, for each horse-power of the engine.

The result of this and all similar rules is, that the steam should always pass through the steam pipe with the same velocity, whatever be the power of the engine.

In engines of the same power, the piston will have very different velocities in the cylinder, according to the effective pressure of the steam, and the proportions and capacity of the cylinder. It is clear from what has been already explained, that, when the power is the same, the same actual quantity of water, in the form of steam, must pass through the cylinder per minute; but, if the steam be used with a considerable pressure, being in a condensed state, the same weight of it will occupy a less space ; and consequently the cylinders of high-pressure engines are smaller than those of the same power in low-pressure engines : the magnitude of the cylinder and the piston therefore, as well as the velocity of the latter, will depend first upon the pressure of the steam.

But with steam of a given pressure, the velocity of the piston will be different: with a given capacity of cylinder, and a given pressure of steam, the power of the engine will determine the number of strokes per minute. But the actual velocity of the piston will depend, in that case, on the proportion which the diameter of the cylinder bears to its length ; the greater the diameter of the piston is with respect to its length, the less will be its velocity. In case of stationary engines used on land, that proportion of the diameter of the cylinder to its length is selected which is thought to contribute to the most efficient performance of the machine. According to some engineers, the length of the cylinder should be twice its diameter ;* others make the length equal to two diameters and a half; but there are circumstances in which considerations of practical convenience render it necessary to depart from these proportions. In marine engines, where great length of cylinder would be inadmissible, and where, on the other hand, considerable power is required, cylinders of short stroke and great diameter are used. In these engines the length of the stroke is often not greater than the diameter of the piston, and sometimes even less.

The actual velocity which has been found to have the best practical effect, for the piston in low-pressure engines, is about 200 feet per minute. This, however, is subject to some variation.

(140.) A given weight or measure of fuel burnt under the

* This is the proportion under which the cylinder with a given capacity will present the least possible surface to the cooling effect of the atmosphere.

boiler of an engine is capable of producing a mechanical effect through the means of that engine, which, when expressed in an equivalent number of pounds' weight lifted a foot high, is called the *duty* of the engine. If all the heat developed in the combustion of the fuel could be imparted to the water in the boiler, and could be rendered instrumental in producing its evaporation ; and if, besides, the steam thus produced could be all rendered mechanically available at the working point ; then the duty of the engine would be the entire undiminished effect of the heat of combustion ; but it is evident that this can never practically be the case. In the first place, the heat developed by the combustion can never be wholly imparted to the water in the boiler : some part of it will necessarily escape without reaching the boiler at all ; another portion will be consumed in heating the metal of the boiler, and in supplying the loss by radiation from its surface ; another portion will be abstracted by the various sources of the waste and leakage of steam ; another portion will be abstracted by the reaction of the condensed steam ; and another portion of the power will be consumed in overcoming the friction and resistance of the engine itself. It is apparent that all these sources of waste will vary according to the circumstances and conditions of the machine, and according to the form and construction of the furnace, flues, boilers, &c. The duty, therefore, of different engines will be different ; and when such machines are compared, with a view to ascertain their economy of fuel, it has been found necessary carefully to register and to compare the fuel consumed with the weight or resistance overcome. In engines applied to manufactures generally, or navigation, it is not easy to measure the amount of resistance which the engine encounters, but when the engine is applied to the pumping of water, its performance is more easily determined.

In the year 1811, several of the proprietors of the mines n Cornwall, suspecting that some of their engines might not be doing a duty adequate to their consumption of fuel, came to a determination to establish a uniform method of testing the performance of their engines. For this purpose a counter was attached to each engine, to register the number of strokes of the piston. All the engines were put under the superintendence of Messrs. Thomas and John Lean, engineers ; and the different proprietors of the mines, as well as their directing engineers, respectively pledged themselves to give every facility and assistance in their power for the attainment of so desirable an end. Messrs. Lean were directed to publish a monthly report of the performance of each engine, specifying the name of the mine, the size of the cylinder, the load upon the engine, the length of the stroke, the number of pump lifts, the depth of the lift, the diameter of the pumps, the time worked, the consumption of coals, the load on the pump, and, finally, the duty of the engine, or the number of pounds lifted one foot high by a bushel of coals. The publication of these monthly reports commenced in August, 1811, and have been regularly continued to the present time.

The favourable effect which these reports have produced upon the vigilance of the several engineers, and the emulation they have excited, both among engine-makers and those to whom the working of the machines are intrusted, are rendered conspicuous in the improvement which has gradually taken place in the performance of the engines, up to the present time. In a report published in December, 1826, the highest duty was that of an engine at Wheal Hope mine in Cornwall. By the consumption of one bushel of coals, this engine raised 46,838,246 pounds a foot high, or, in round numbers, forty-seven millions of pounds.

In a report published in the course of the present year (1835) it was announced that a steam engine, erected at a copper mine near St. Anstell, in Cornwall, had raised by its average work 95 millions of pounds 1 foot high, with a bushel of coals. This enormous mechanical effect having given rise to some doubts as to the correctness of the experiments on which the report was founded, it was agreed that another trial should be made in the presence of a number of competent and disinterested witnesses. This trial accordingly took place a short time since, and was witnessed by a number of the most experienced mining engineers and agents: the result was, that for every bushel of coal consumed under the boiler the engine raised $125\frac{1}{2}$ millions of pounds weight one foot high.

(141.) It may not be uninteresting to illustrate the amount of mechanical virtue, which is thus proved to reside in coals, in a more familiar manner.

Since a bushel of coal weighs 84 lbs. and can lift 56,027 tons a foot high, it follows that a pound of coal would raise 667 tons the same height; and that an ounce of coal would raise 42 tons one foot high, or it would raise 18 lbs. a mile high.

Since a force of 18 lbs. is capable of drawing 2 tons upon a railway, it follows that an ounce of coal possesses mechanical virtue sufficient to draw 2 tons a mile, or 1 ton 2 miles, upon a level railway.*

The circumference of the earth measures 25,000 miles. If it were begirt by an iron railway, a load of one ton would be drawn round it in six weeks by the amount of mechanical power which resides in the third part of a ton of coals.

The great pyramid of Egypt stands upon a base measuring 700 feet each way, and is 500 feet high; its weight being 12,760,000,000 lbs. To construct it, cost the labour of 100,000 men for 20 years. Its materials would be raised from the ground to their present position by the combustion of 479 tons of coals.

The weight of metal in the Menai bridge is 4,000,000 lbs.,

* The actual consumption of coal upon railways is in practice about eight ounces per ton per mile. It is, therefore, worked with sixteen times less effect than in the engine above-mentioned.

and its height above the level of the water is 120 feet: its mass might be lifted from the level of the water to its present position by the combustion of 4 bushels of coals.*

The enormous consumption of coals in the arts and manufactures, and in steam navigation, has of late years excited the fears of some persons as to the possibility of the exhaustion of our mines. These apprehensions, however, may be allayed by the assurance received from the highest mining and geological authorities, that, estimating the present demand from our coal mines at 16 millions of tons annually, the coal fields of Northumberland and Durham alone are sufficient to supply it for 1700 years, and after the expiration of that time the great coal basin of South Wales will be sufficient to supply the same demand for 2000 years longer.

But, in speculations like these, the probable, if not certain, progress of improvement and discovery ought not to be overlooked; and we may safely pronounce that, long before a minute fraction of such a period of time shall have rolled over, other and more powerful mechanical agents will altogether supersede the use of coal. Philosophy already directs her finger at sources of inexhaustible power in the phenomena of electricity and magnetism. The alternate decomposition and recomposition of water, by magnetism and electricity, has too close an analogy to the alternate processes of vaporization and condensation, not to occur at once to every mind: the developement of the gases from solid matter by the operation of the chymical affinities, and their subsequent condensation into the liquid form, has already been essayed as a source of power. In a word, the general state of physical science at the present moment, the vigour, activity, and sagacity with which researches in it are prosecuted in every civilized country, the increasing consideration in which

* Some of these examples were given by Sir John Herschel, in his Preliminary Discourse on Natural Philosophy; but since that work was written an increased power has been obtained from coals, in the proportion of 7 to 12½.

scientific men are held, and the personal honours and rewards which begin to be conferred upon them, all justify the expectation that we are on the eve of mechanical discoveries still greater than any which have yet appeared; and that the steam engine itself, with the gigantic powers conferred upon it by the immortal Watt, will dwindle into insignificance in comparison with the hidden powers of nature still to be revealed; and that the day will come when that machine, which is now extending the blessings of civilization to the most remote skirts of the globe, will cease to have existence except in the page of history.

---

## CHAPTER XIX.

### PLAIN RULES FOR RAILWAY SPECULATORS.

(142.) For some time after the completion of the Liverpool and Manchester railway, doubts were entertained of its ultimate success as a commercial speculation; and, even still, after several years' continuance, some persons are found, skeptical by temperament, who have not acquired full confidence in the permanency of its advantages. The possibility of sustaining a system of regular transport upon it, with the unheard-of speed effected at the commencement of the undertaking, was, for a long period, questioned by a considerable portion even of the scientific world; and, after that possibility was established, by the regular performance of some years, the practicability of permanently profitable work, at that rate of speed, was still doubted by many, and altogether denied by some. The numerous difficulties to be encountered, and the enormous expense of locomotive power, have been fully admitted by the directors in their semi-annual reports. Persons interested in canals and other rival establishments, and others constitutionally doubtful of every thing, attributed the dividends to the indirect proceedings of the managers, and asserted that when they appeared to be sharing their profits, they, in reality, were sharing their capital. This delusion, however, could not long continue, and the payment of a steady semi-annual divided of $4\frac{1}{2}$ per cent. since the opening of the railway, together with the commencement of a reserved fund of a considerable amount, with a premium of above 100 per cent. on the original shares, has brought conviction to understandings impenetrable to general reasoning ; and the tide of opinion, which, for a time, had turned against railways, has now, by the usual reaction, set in so violently in their favour, that it becomes the duty of those who professionally devote themselves to such inquiries, to restrain and keep within moderate bounds the public ardour, rather than to stimulate it.

The projects for the construction of great lines of internal communication which have been announced would require, if realized, a very large amount of capital. Considering that the estimated capital is invariably less than the amount actually required, we shall not, perhaps, overrate the extent of the projected investments if we estimate them at fifty millions. The magnitude of this amount has created alarm in the minds of some persons, lest a change of investment so extensive should produce a serious commercial shock. It should, however, be considered that, even if all the projected undertakings should be ultimately carried into execution, a long period must elapse, perhaps not less than fifteen or twenty years, before they can be all completed : the capital will be required, not suddenly, but by small instalments, at distant intervals of time. Even if it were true, therefore, that, to sustain their enterprises, an equivalent amount of capital must be withdrawn from other investments, the transfer would take place by such slow degrees as to create no serious inconvenience. But, in fact, it is not probable that any transfer of capital whatever will be necessary. Trade and manufactures are at the present moment in a highly flourishing condition; and the annual accumulation of capital in the country is so great, that the difficulty will probably be, not to find capital to meet investments, but to find suitable investments for the increasing capital. In Manchester alone, it is said that the annual increment on capital is no less than three millions. In fifteen years, therefore, this mart alone would be sufficient to supply all the funds necessary for the completion of all the proposed railroads, without withdrawing capital from any other investment.

The facilities which these Joint Stock Companies offer for the investment of capital, even of the smallest amount, the temptations which the prospect of large profits hold out, and the low interest obtained on national stock of every description, have attracted a vast body of capitalists, small and great, who have subscribed to these undertakings with the real intention of investment. But, on the other hand, there is a very extensive body of speculators who engage in them upon a large scale, without the most distant intention, and, indeed, without the ability, of paying up the amount of their shares. The loss which the latter class of persons may sustain would, probably, excite little commiseration, were it not for the consequences which must result to the former, should a revolution take place, and the market be inundated with the shares of these gambling speculators, who buy only to sell again. Effects would be produced which must be ruinous to a large proportion of the *bonâ fide* subscribers. It may, therefore, be attended with some advantage to persons who really intend to make permanent investments of this nature, to state, in succinct and intelligible terms, the principal circumstances on which the efficiency and economy of railroads depend, so as to enable them, in some measure, to form a probable conjecture of the prospective advantages which the various projects hold out. In doing this we shall endeavour, as much as possible, to confine our statements to simple facts and results, which can neither be denied nor disputed, leaving, for the most part, the inferences to which they lead to be deduced from them by others.

It may be premised, that persons proposing to engage in any railroad speculation should obtain *first* a table of *gradients;* that is, an account of all the acclivities upon the line from terminus to terminus, stating how many feet in a mile each incline rises or falls, and its length. *Secondly,* it would also be advantageous to have a statement of the lengths of the radii of the different *curves,* as well as the lengths of the curves themselves. *Thirdly,* an account of the actual intercourse which has taken place, for a given time, upon the turnpike road connecting the proposed termini, stating the number of coaches licensed, and the average number of passengers they carry; also as near an account of the transport of merchandise as may be obtained. The latter, however, is of less moment. An approximate estimate may be made of the intercourse in passengers, by allowing for each coach, upon each trip, half its licensed complement of load. *Fourthly,* the water communication by canal or otherwise between the places; and the amount of tonnage transported by it. With the information thus obtained, the following succinct maxims will be found useful:—

I.

No railroad can be profitably worked without a large intercourse of passengers. Goods, merchandise, agricultural produce, &c. ought to be regarded as of secondary importance.

II.

A probable estimate of the number of passengers to be expected upon a projected line of railroad may be made by increasing the average number of passengers for the last three years, by the common road, in a twofold proportion.

The average number of passengers daily between Liverpool and Manchester, before the formation of the railway, was about 450; the present average number is above 1300. A short railroad of about five miles is constructed between Dublin and Kingstown: on which the average number of passengers daily between those places has increased in nearly the same proportion.

### III.

Passengers can be profitably transported by canal, at a speed not exceeding nine miles an hour, exclusive of delays at locks, at the rate of one penny per head per mile. The average fares charged upon the Manchester railway are at the rate of $1\frac{84}{100}d.$ per head per mile, the average speed being twenty miles an hour.

To transport passengers at the rate of ten miles an hour on a railway would cost very little less than the greater speed of twenty miles an hour, so that a railroad could not enter into competition on equal terms with a canal by equalising the speed.

The canal between Kendal and Preston measures 57 miles: passengers are transported upon it between these places at the average speed of a mile in $6\frac{1}{2}$ minutes, or $9\frac{1}{4}$ miles an hour nearly, exclusive of delays at locks. The fare charged is at the average rate of a penny a mile. There are eight locks, rising 9 feet each, and a tunnel 400 yards long, through which the boat is tracked by hand; the tunnel requires 5 minutes, and the locks from 25 to 28 minutes, in descending, and 45 to 48 minutes in ascending.

Similar boats are worked on the Forth and Clyde, and the Union Canals in Scotland, and on the Paisley and Johnstone canal, at nearly the same fares.

IV.

At the fare of $1\frac{84}{160}d.$ per head per mile, the profit on the Manchester railroad is 100 per cent. on the disbursements for passengers.

V.

Goods can be profitably transported by canal at a lower tonnage than by railroads; the speed on the canal (for goods) being, however, but one-fifth of the speed on the railroad.

VI.

Goods are transported on the Liverpool and Manchester railroad at three-pence three farthings per ton per mile, with a profit of about 40 per cent. upon the disbursement, having the competition of a canal between its termini.

VII.

A long railroad can be worked with greater relative economy than a short one.

VIII.

Steam engines work with the greatest efficiency and economy, when the resistance they have to overcome is perfectly uniform and invariable.

IX.

The variation of resistance on railroads depends, first, on acclivities; secondly, on curves.

By curves are meant the changes of direction of the road to the right or to the left. The direction of a railroad cannot be changed suddenly by an angle, but must be effected gradually by a curve. Supposing the curve to be (as it generally is) the arc of a circle, the radius of the curve is the distance of the centre of the circle from the curve. This radius is an important element in the estimate of the road.

X.

The more nearly a railroad approaches to an absolute level, and perfect straightness, the more profitably will it be worked.

XI.

The total amount of mechanical power necessary to transfer a given load from one extremity of a railroad to another is a matter of easy and exact calculation, when the gradients and curves are known; and the merits of different lines may be compared together in this respect: but it is not the only test of their efficiency which must be applied.

XII.

A railroad having gradients exceeding seventeen feet in a mile will require more mechanical power to work it than it would were it level; and the more of these excessive gradients there are upon it, and the more steep they are, the greater will be this disadvantage.

XIII.

Although a railroad having no gradients exceeding seventeen feet in a mile does not require more mechanical power than a level, yet the mechanical power which it requires will not be so advantageously expended, and, therefore, it will not be so economical.

XIV.

A railroad which has gradients above thirty feet in a mile will require such gradients to be worked by assistant locomotive engines, which will be attended with a waste of power, and an increase of expenditure, more or less, according to the number and length of such gradients.

XV.

A very long inclined plane cannot be worked by an assistant locomotive without a wasteful expense. Gradients exceeding seventeen feet per mile must, therefore, be short.

XVI.

Gradients exceeding fifty feet in a mile cannot be profitably worked except by stationary engines and ropes, an expedient attended with so many objections as to be scarcely compatible with a large intercourse of passengers.

XVII.

Steep gradients, provided they descend from the extremities of a line, are admissible provided they be short.

It is evident that in this case the inclined planes will help at starting to put the trains in motion, at the time when, in general, there would be the greatest strain upon the moving power; and, in approaching the terminus, the momentum would be sufficient to carry the train to the top of the plane, if its length were not great, since it must, at all events, come to a stop at the extremity.

XVIII.

The effect of gradients in increasing the resistance during the ascent may be estimated by considering that a gradient of seventeen feet in a mile doubles the resistance of the level, thirty-four feet in a mile triples it, and eight and a half feet in a mile adds one half its amount, and so on.

XIX.

With the speed now attainable on railways, curves should be avoided with radii shorter than a mile. Expedients may diminish the resistance, but, through the negligence of engine drivers, they must always be attended with danger. Curves are not objectionable near the extremities of a line.

XX.

The worst position for a curve is the foot of an inclined plane, because of the velocity which the trains acquire in the descent, and the occasional impracticability of checking them.

XXI.

In proportion as the speed of locomotives is increased by the improvements they are likely to receive, the objections and dangers incident to curves will be increased.

The difficulty which attends the use of long tunnels arises from the destruction of the vital air which is produced by the combustion in the furnaces of the engines. Tunnels on a level should, therefore, be from twenty-five to thirty feet high, and should be ventilated by shafts or other contrivances.

XXIII.

The transition from light to darkness, the sensation of humidity, and the change in summer from a warm atmosphere to a cold one, will always form an objection to long tunnels on lines of railroad intended for a large intercourse of passengers.

XXIV.

All the objections to a tunnel are aggravated when it happens to be upon an acclivity. The destruction of vital air in ascending it will be increased in exactly the same proportion as the moving power is increased. Thus, if it ascend 17 feet in a mile, the destruction of vital air will be twice as great as on a level; if it ascend 34 feet in a mile, it will be three times as great; 51 feet in a mile, four times as great, and so on.

XXV.

If by an overruling necessity a tunnel is constructed on an acclivity, its magnitude and means of ventilation should be greater than on a level, in the same proportion as the resistance produced by the acclivity is greater than the resistance upon a level.

XXVI.

Tunnels should be ventilated by shafts at intervals of not more than 200 yards.

XXVII.

While a train is passing through a tunnel, no beneficial ventilation can be obtained from shafts. The engine will leave behind it the impure air which it produces, and the passen gers will be enveloped in it before it has time to ascend the shafts. Sufficient magnitude, however, may be given to the tunnel to prevent any injurious consequences from this cause. A disagreeable and inconvenient odour will be experienced.

XXVIII.

Tunnels on a level, the length of which do not exceed a third of a mile, will probably not be objectionable. Tunnels of equal length upon acclivities would be more objection able.

I may observe generally that we have as yet little or no experience of the effect of tunnels on lines of railroad worked by locomotive engines, where there is a large intercourse of passengers. On the Leicester and Swannington railroad, there is a tunnel of about a mile long, on a part of the road which is nearly level ; it is ventilated by eight shafts, and I have frequently passed through it with a locomotive engine. Even when shut up in a close carriage the annoyance is very great, and such as would never be tolerated on a line of road having a large intercourse in passengers. This railroad is chiefly used to take coals from some collieries near Swannington, and there is no intercourse in passengers upon it, except of the labouring classes from the adjacent villages: the engines burn coal, and not coke ; and they consequently produce smoke, which is more disagreeable than the gases which result from the combustion of coke. This tunnel also is of small calibre.

On the Leeds and Selby railroad there is a tunnel, on a part which is nearly level, the length of which is 700 yards, width 22 feet, height 17. It is ventilated by three shafts of about 10 feet diameter and 60 feet high. There is an intercourse of passengers amounting to four hundred per day upon this road, and, generally speaking, they do not object to go through the tunnel with a locomotive engine. The fuel is coke.

(*l*) In order to show the present state of railroad transportation in the United States, and enable our readers to compare it with the opinions and facts adduced by Dr. Lardner, we take the latest accounts from the Charleston and Hamburgh Railroad. The engines drag a train of cars which carry a load of 130 tons, and perform the distance (240 miles) in three days, travelling only by daylight. With these loads they mount planes having inclinations of 37 feet per mile. The same engines are capable of carrying passengers at the rate of 40 miles per hour, and often perform 30, but their average speed is limited by regulation to 20 miles per hour.

This railroad is remarkable for being the largest which has yet been constructed, and is besides an object of just pride, inasmuch as it was commenced at a time, when, according to Dr. Lardner, the subject was but imperfectly understood even in Europe, and all its arrangements are due to native talent and skill, unassisted by previous discoveries in Europe.—A. E.

# INDEX.

## A.



## B.



## C.

## K.



## L.



## M.



## N.



## O.

## P.



## R.



## S.

## T.



## U

## V.



## W.



THE END.

*Bartlett's Natural Philosophy for Colleges*

# ELEMENTS OF NATURAL PHILOSOPHY.

IN THREE VOLUMES.

BY W. H. C. BARTLETT, LL.D.,

*Professor of Natural and Experimental Philosophy in the United States Military Academy at West Point.*

## Vol. 1st.—MECHANICS.

The present volume is the first of three, in which its author desires to offer to Academies and Colleges a Course of Natural Philosophy, including Astronomy. It embraces the subject of Mechanics—the groundwork of the whole. It is intended to be complete within itself, and to have no necessary dependence for the full comprehension of its contents upon those which are to follow.

---

"*Rensselaer Polytechnic Institute, Troy, N. Y., November* 1, 1850.

"I congratulate Professor Bartlett on the result of his labors, in the production of the present treatise, on a subject so eminently important, whether in its general relation to a system of liberal culture, or, in a more technical sense, as a foundation of the science and art of construction.

"With the somewhat hasty examination which I have been, as yet, enabled to give it, I cannot presume to indicate specialities; but I am abundantly satisfied of its vast advance, in every respect, upon any *American* predecessor; and it is a consideration not a little gratifying, that we are now enabled to present our students with a manual which shall exhibit the most important principles of mechanical science, in the garb of modern analysis, vigorously demonstrated and fairly illustrated, with a fullness that appears to leave little to be desired.

"We shall immediately adopt the work in our institution.

"B. FRANKLIN GREENE,
"Director and Professor of Mechanics, etc.,
"Rensselaer Polytechnic Institute, Troy, N. Y."

---

"The work is intended for the use of students in our highest colleges and institutions of learning, and is much more complete and serviceable than any thing of the kind hitherto published. Professor Bartlett is well known as one of the most thoroughly scientific men in the United States; and the success with which he has presided over this department of instruction in the West Point Military Academy for many years, affords decisive evidence of his abundant and pre-eminent qualifications for the preparation of such a work. It is published in a large and very elegantly printed octavo volume, and is very profusely illustrated with diagrams. It is a work which cannot fail, as rapidly as it becomes known, to find its way into all our colleges in which completeness and thoroughness of scholarship are properly prized."—*Courier and Enquirer.*

---

"The progress of research and demonstration constantly going on, requires an occasional recast of our text-books; and we have every reason in Prof. Bartlett's attainments, position, and facilities, for believing that the work he has so elaborately undertaken will answer the requisitions of the present advanced state of the science, as well as of the improved method of study. The present volume treats of mechanics, both of solids and fluids. An introductory chapter defines the objects of the science, and the properties of bodies. The author possesses a concise and unambiguous method, which is well adapted for his purpose; and his demonstrations, embracing both analytical and synthetic processes, are very concise and clear. It is evident that he has possessed himself of all the learning on his topics extant, and claims to have embodied, especially, the best results of the labors of two eminent French philosophers—M. Poncellet and M. Peschel."—*New York Evangelist*

*Bartlett's Elements of Natural Philosophy.*

# ACOUSTICS AND OPTICS.

## BY W. H. C. BARTLETT, LL.D.,

PROFESSOR OF NATURAL AND EXPERIMENTAL PHILOSOPHY AT THE UNITED STATES MILITARY ACADEMY, WEST POINT.

VOL. II.

---

The student is here presented with the second volume of a course of natural philosophy intended for the classes in colleges and universities. The work complete will extend to three volumes, of which the first, embracing Mechanics, has already been noticed in the columns of this journal. The book before us treats of Acoustics, including the velocity, divergence, and reflexion of sound; music, chords, intervals, and harmony; and Optics, exhibiting the laws of reflexion and refraction; the telescope, microscope, camera-obscura, and polarization of light. Emanating from the proud seat of science, the Polytechnic School of North America, such works need not the blazonry of praise nor the guerdon of critical encomium. The high standard of pure and mixed science at West Point, and the acknowledged excellence of the Professors, constitute our best guaranty of merit.—*National Intelligencer.*

---

This treatise on two very important branches of Natural Philosophy, namely, Acoustics and Optics, by Professor Bartlett, of West Point, is one of no ordinary character, and is highly creditable, as is his other work on Mechanics, to the talented author. The subjects are well arranged, following as it were, in geometric order; the enunciation is clear and comprehensive; the diagrams abundant, and well executed; the formulæ used in the process of demonstration plain, and nothing introduced but what appertains to the subject. Of all books we have seen published on the science of Optics, the present is the most lucid and instructive; admirably adapted to the present advanced stage of science, and the student, already acquainted with Geometry Trigonometry, and Algebra, can, by means of it, become his own teacher. Schools and colleges cannot adopt a better work on the science of which it treats; and the type, paper, and general appearance of the book, reflects credit on the publishers.

---

ELEMENTS OF NATURAL PHILOSOPHY.—I was gratified to see this addition to our educational appliances in an interesting and important department of physical science. The plan of *generalization* adopted by the author, while in harmony with the teachings and manifest tendencies of modern science, seems also to be peculiarly desirable in a text-book designed to meet the requirements of a sound and judicious instruction.

The idea developed in this treatise, as suggested in the preface of the author, is that of considering the phenomena of Sound and Light from a common point of view, and regarding them as results of precisely similar dynamic conditions of the media in which they respectively originate, of treating them on a basis of certain *fundamental equations alike applicable to both classes of phenomena.*

This idea, although a novelty in an educational treatise of a practical form, is nevertheless, in the *right direction;* and we can scarcely doubt that the present treatise will be found to supply, in a highly satisfactory manner, a *want* frequently experienced among instructors, of a *text-book adapted to more modern, rational, and practical views of instruction on these subjects.* B. F. GREENE, Troy, N. Y.

# ELEMENTS OF ANALYTICAL MECHANICS.

BY WM. H. C. BARTLETT, LL.D.

*Professor of Natural and Experimental Philosophy in the Military Academy at West Point.*

---

This work is intended as a text-book for students who have made considerable progress in the higher branches of Mathematics, the subject of which it treats being discussed analytically, by the aid of the calculus; differing in this respect from the previous and more popular "Elements of Mechanics" by the same author.

---

RENSSELAER POLYTECHNIC INSTITUTE,
Troy, Sept. 13, 1853.

Professor Bartlett's recent effort, the Elements of Analytical Mechanics, will, I am sure, be received as a most welcome addition to the *very small* list of American educational works on the subject on which it treats. In fact, the present treatise is, so far as I am aware, the first attempt, not only in this country, but in the English language, to exhibit the principles of the most important of sciences, in the spirit of that *complete generalization* originally so happily conceived and admirably developed by the genius of a LAGRANGE, in his celebrated *Mécanique Analytique.*

Professor Bartlett, starting with the establishment of a *general* equation, involving the principle of Virtual Velocities, and expressed in a form to include the principle of D'Alembert, develops the whole subject of Statics and Dynamics of Solids and Fluids, by a series of elegant discussions of the fundamental equation.

Such a treatise, while sufficiently comprehensive in scope and compendious in form, seems to be capable, were it generally introduced into our higher institutions, of doing much to elevate the standard as well of Analytical as Mechanical Science among us; since the *utilities* of the former are only truly appreciated and availably grasped by the student, when brought into view in the concrete discussions of Force and Matter.

Yours, &c. B. FRANKLIN GREENE,
*Director of R. P. I., and Prof. of Mechanics and Constructive Engineering.*

---

The substance of this volume has constituted for some years the text-books of the author's class in the United States Military Academy, and is now for the first time published. The classification adopted—the result of much thought—arranges the subject under the heads of Mechanics of Solids and Mechanics of Fluids; together with a third part, embracing the application of the principles evolved under the two former heads.

The honorable position occupied by Prof. Bartlett, together with the merit of his contributions to science, will insure a favorable reception for this new treatise. Seldom, indeed, have we met a more beautiful condensation of expression than that employed to describe the scope of this branch of mixed mathematics.—*National Intelligencer.*

www.ingramcontent.com/pod-product-compliance
Lightning Source LLC
LaVergne TN
LVHW010200110826
845151LV00002B/569

* 9 7 8 1 4 2 5 5 3 7 5 5 5 *